Next-Generation Textiles

Edited by Hassan Ibrahim

Published in London, United Kingdom

Next-Generation Textiles
http://dx.doi.org/10.5772/intechopen.97899
Edited by Hassan Ibrahim

Contributors
Wei Tiancheng Wei, Yu Sun, Eunkyoung Shim, Ramratan Guru, Anupam Kumar, Rohit Kumar, Deepika Grewal, Fulya Yilmaz, Gamze D. Tetik, Gizem Celep, Jerry Ochola, Michele Conti, Hassan Ibrahim, Ahmed Amr

First published in London, United Kingdom, 2023 by IntechOpen
IntechOpen is the global imprint of INTECHOPEN LIMITED, registered in England and Wales, registration number: 11086078, 5 Princes Gate Court, London, SW7 2QJ, United Kingdom

British Library Cataloguing-in-Publication Data
A catalogue record for this book is available from the British Library

Additional hard and PDF copies can be obtained from orders@intechopen.com

Next-Generation Textiles
Edited by Hassan Ibrahim
p. cm.
Print ISBN 978-1-80355-882-0
Online ISBN 978-1-80355-883-7
eBook (PDF) ISBN 978-1-80355-884-4

Meet the editor

Dr. Hassan Ibrahim is a professor at the Textile Research and Technology Institute, National Research Center, Cairo, Egypt. He has expertise in applied chemistry and technology of organic chemistry, especially in medical textiles, carbohydrates chemistry and technology, and polymers. He is also interested in pollution prevention, preparation, and applications of nanoparticles and nanofibers (polymer chemistry, chemistry of chitosan, chitosan modification, nanoparticle preparation, and electrospinning). He has participated in several national and international projects. He is also a member of and the national representative of Egypt for the International Union for Pure and Applied Chemistry (IUPAC) Division VI. Professor Ibrahim has supervised three Ph.D. and twelve MSc students.

Contents

Preface **IX**

Section 1
Medical Textiles **1**

Chapter 1 **3**
Study of the Implantable and Non-Implantable Application in Medical Textile
by Ramratan Guru, Anupam Kumar, Deepika Grewal and Rohit Kumar

Chapter 2 **17**
Healthcare and Hygiene Products Application in Medical Textile
by Ramratan Guru, Anupam Kumar and Rohit Kumar

Chapter 3 **29**
Bacterial Cellulose: Biosynthesis and Applications
by Ahmed Amr and Hassan Ibrahim

Section 2
Recycled Textiles **53**

Chapter 4 **55**
Progress of Recycled Polyester in Rheological Performance in Molding, and Economic Analysis of Recycled Fibers in Fashion and Textile Industry
by Wei Tiancheng Wei, Yu Sun and Eunkyoung Shim

Chapter 5 **71**
Limitations of Textile Recycling: The Reason behind the Development of Alternative Sustainable Fibers
by Gizem Celep, Gamze D. Tetik and Fulya Yilmaz

Section 3
Textile Investigation **97**

Chapter 6 **99**
Numerical Investigation of Braided Structure Potential as a Cast for Femur Shaft Fracture
by Jerry Ochola and Michele Conti

Chapter 7 113
A Review of Significant Advances in Areca Fiber Composites
by Narayanan Gokarneshan, Venkatesan Sathya, Jayagopal Lavanya, Shaistha Shabnum, Habeebunisa and Sona M. Anton

Preface

Intelligent textiles represent the next generation of fibers and fabrics and are an exciting new area of research. They have numerous applications not only in functional fashion but also in health care and hygiene. Their development came about due to recent advancements such as areca fiber-reinforced composites and developments in bacterial cellulose.

In various chapters, this book discusses the importance of knits in the applications of next-generation textiles. Chapters include images and present recent studies in the field of next-generation intelligent textiles that show promise for treating diseases affecting the body's key organs. They also point to areas of application where more research is possible. The recently created hygiene and healthcare items made of sustainable fibers have also been assessed and discussed. The role of textile recycling restrictions has been covered as well in one chapter. The advancements in implantable and non-implantable medical textile applications are covered in a number of chapters.

I would like to express my gratitude to the researchers whose significant achievements are highlighted in this book, and I am convinced that it will inspire others to conduct in-depth studies in the field. As a result, the book makes for engaging reading and is a useful resource for scholars of textiles and medicine.

Hassan Ibrahim
Pre-Treatment and Finishing of Cellulosic Fibers,
Textile Research and Technology Institute,
National Research Centre,
Cairo, Egypt

Section 1

Medical Textiles

Chapter 1

Study of the Implantable and Non-Implantable Application in Medical Textile

Ramratan Guru, Anupam Kumar, Deepika Grewal and Rohit Kumar

Abstract

Nowadays medical textiles are one of the more continuous growing parts in technical textile market. The generally medical textile should have strength, biodegraded, nontoxic, biologically compatible, dimensional stability, resistant to allergens and cancer, more comfort human body, antifungal and antimicrobial performance. Development with inside the discipline of textiles, either natural or manmade textiles, typically aimed toward how they beautify the consolation to the users. Development of medical textiles may be taken into consideration as one such development, that's virtually supposed for changing the painful days of sufferers into the snug days. The basically are used the implantable materials to repair the affected parts of the person body. The generally are used in wound sutures and used surgery time replacement and other segment to replacement like artificial ligaments, vascular grafts. This includes type of the sutures, soft tissue implants, orthopedic implants, cardiovascular implants etc. Non-implantable materials are used for external applications for role of bandages, wound care and wound care products, plasters etc. This paper are discusses the main role of implantable and non-implantable medical textile products.

Keywords: medical textile, design parameters, implantable, non-implantable application

1. Introduction

Some desirable properties of medical fibers include non-toxicity strength ability, biocompatibility, biodegradability, good absorbability, softness and freedom from additives and contaminates. The textile material and scientification technics has used generally in medical, surgical application like strength, flexibility, comfort and antimicrobial performances. The basically medical material products are made to multifilament and monofilament yarn, these are made by knitted, nonwoven, woven, braided fabrics and composite structures [1]. The term medical textile literally means textile used for medical purposes. Newsday around the world in textile industries are more growing part of the medical sectors and hygiene products. Medical textiles represent one of the maximum dynamic studies fields' features of technical textiles and its variety of applications. They constitute systems designed and done for a scientific application (intra body/greater body, implantable and non-implantable) textiles utilized in organic structures to estimate, treat, growth or regenerate a tissue, organ or characteristic of the body (plaster, dressings, bandages, strain garments) [2].

Absorbency, high flexibility, softness, high strength, non-toxicity and biocompatibility of textile materials are the key factors which has fuelled the growth of the textiles for its use in implantable, non-implantable, extracorporeal and hygienic products[1]. Although the natural way to replace a defective body part is the transplantation method, however owing to a number of incentives counting availability this is not always possible thus implantable textiles in the form of fiber and fabric are used in effective repair to the body. Sutures, soft tissue implants, orthopedic and cardiovascular grafting are the implantable textiles which has helped medical science in achieving unparalleled success in recent times [3, 4]. Non-implantable substances are utilized in outside packages, which can also additionally or might not keep in touch with the skin. The substances used must be nonallergenic, anti-cancer, anti-bacterial, permeable to air have a very good capacity to take in liquids, excessive capillarity and wettability, permit moisture shipping and feature the capacity to be sterilized. The foremost packages of those substances confer with wound care and bandages. These materials can be classified into two separated and specialization areas of application. Implantable materials: sutures or wound closure, vascular grafts, artificial ligaments, artificial joints. Non-implantable materials: wound dressing, bandages, plasters, pressure garments, orthopedic belts etc.

2. Implantable textiles

These are used for replacing diseased organ or tissue within the body. These replacements must be non-toxic and biocompatible. The implants are normally used for replacing arteries, heart valves, joints etc. Two types of fibers are used for implantable textile.

2.1 Biodegradable fibers

These are the fibers which are degraded by biological conditions within 2–3 months and mostly used inside the body. These include collagen, alginate, polyactide, polyglycolide, polyamine and some polyurethane [5, 6].

2.2 Non-biodegradable fibers

- These are the fibers which are not degraded by biological condition for a long time and mostly used for external purposes. These include polytetrafluoroethylene (PTFE), polyester, polypropylene, carbon and others.

- Factors which are important for implantable textiles are:

- Biocompatibility and biostability

- The properties of polyester will influence the success of implantation in terms of biodegradability (**Tables 1** and **2**).

Parameters	Fiber type	Fabric type
Cardiovascular implants, vascular grafts, heart valves	Polyester, PTFE	Knitted, woven
Artificial tendons, artificial ligaments, artificial skin, artificial lumen, eye contact lenses etc.	PTFE, polyester, polyamide, silk, carbon, etc.	Woven, braided
Sutures thread	Collagen, polylactide, polyglycolide, polyester, polyamide, PTFE, polypropylene, polyethylene	Mono-filament, braided
Orthopedic implants, artificial joints, artificial bones etc.	Silicone, polyacetal, polyethylene, polysulphone, carbon, polyester, glass, ceramic	Composite

Table 1.
The implantable ingredients application in medical sectors [1–4, 15].

Application	Implant
Abdominal wall, hernia	Meshes, patches
Blood vessel	Tubular prostheses (woven, knitted, nonwoven), stents, stent graft coatings
Dura	Patches (nonwoven)
Heart	Patches, occluder, suturing ring of valves
Osteosynthesis	Fiber reinforced devices, cords for fixation
Tendon/ligament	Reinforcement
Trachea, esophagus	Prostheses

Table 2.
Major applications for implantable textile medical devices [5–8, 15].

3. Sutures

Suture is a generic term for all materials used to bring the served body tissue together and to hold these tissues in their normal position until healing takes place. Sutures are threads that are used as the way of repairing damaged tissues, cut vessels and surgical incisions by uniting the basic edges of the wounds in their required sites. It provides the necessary strength and a temporary barrier to prevent the unwanted infection. The key qualities stimulating the suture design are universal applicability, easy to handle, no kinks, coiling, twisting, or levitating, biocompatibility, inertness, uniformity in tensile strength in terms of suture type and size, frictionless surface to glide through tissue high friction for secure knotting, sterilizable without composition changes, complete absorption i.e. no residue after healing. A suture is a thread that both approximated and maintains tissues until the natural healing process has provided a sufficient level of wound strength or compresses blood vessels in order to stop bleeding. Sutures for wound closure are either monofilament or multifilament threads twisted, spun together or braided. They can also be dyed, undyed, coated or uncoated [7]. Patients' safety is major factor for application of a suture. An incision into the lung would need to be closed using a suture with a high elasticity level, slow degradation rate and high tensile strength level. So, a surgery is never successful if the wound is not sutured or closed in a proper manner as to promote healing in a timely and safe fashion also if the suture of a rough morphology (e.g. braided) the tissue will swell more and more susceptible to infection than if a smooth suture (e.g. monofilament) is used [8].

3.1 Classification and types of sutures

The classification of the sutures may be done as follows into two types depending on their nature and structure:

3.1.1 Assimilated type (absorbable sutures)

Assimilated type of sutures is intended to be absorbed by the body i.e. to be broken down in the body and a second surgery for their removal is not desired. e.g. catgut, collagen and poly glycolic acid. Catgut is one of the most commonly used materials for the manufacture of sutures and is extracted from the ox bone. Being highly absorbable it can also be implanted in the human body even in the case of an infection however its strength deteriorates to half after a week in the body, regardless of the fact that 3 weeks are required for the recovery of an incision after surgery [9].

3.1.2 Non-assimilated type (non-absorbable)

Non-assimilated types of sutures are considered to be implanted for long term and need to be removed latter. (E.g. cotton, silk, polyester, polyamide and polyethylene.) Cotton sutures necessitate meticulous aseptic technique during use. The main benefit of such sutures is that they are not irritant and the shortcoming is that it is the weakest suture material. Despite the possession of necessary physical form, compatibility and mechanical properties, the very slow biodegradation of the silk filament and the need for the surgical removal is the main draw back in many applications (**Figure 1**) [10].

The different types of suture include monofilament suture, a braided suture, a pseudo monofilament suture and a twisted strand suture each having its own positive and negative points. Monofilament sutures are made of a single filament of polyester, polyamide, polypropylene or polydioxanone and offer smooth suture drag and low tissue drag. Using such sutures, it is easy to make or place a knot in the depth of the body

Figure 1.
Nylon monofilament suture [9, 10].

although the security and the flexibility of the knot are low. In braided type of sutures 8–16 polyester, polyamide or silk monofilaments are braided and coated with a lubricant to increase the flexibility and handle of the sutures. A pseudo monofilament sutures have a core of several twisted materials coated with an extrusion of the same material. It offers low tissue drag, good knottability, low knot security and fair flexibility.

3.1.3 Intelligent sutures

The basically are used sutures in the surgical operation and other injuries. The basically are used suture thread length to tie blood vessels or sew tissues part of body. The many types of suture threads are used as absorbable performance characteristics. All this absorbable intelligent materials technique in sutures are very good working and this is doing better performance in medical sectors. This types all material are used biodegradable and biocompatible polymer. The generally many types of absorbable suture are used made from synthetic polymers.

4. Soft tissue implants

The soft tissues are utilization in biomedical materials application like artificial tender, artificial corners and artificial prosthness etc. There are two main thrust of tissue engineering research. They are (i) the in vivo route and (ii) the in vitro approach. The objective of in vivo route is to initiate tissue engineering therapies inside the body for the repair and regeneration of damaged or diseased tissue. This approach can be successful for blood cell and nerve regeneration (both peripherial and spiral cord), skin repair, remodeling of defective bone, cornea and retina and for repairing damaged myocardium (heart muscle) following a myocardial infarction (heart attack). Not all diseases and injuries can be controlled by in vivo therapies. For example use in complex tissue cultures for the production of enzymes, drug and growth factors and for toxicological and pharmacological assays. It is depending on the medical sectors application. Ligament implants are carried out to provide autologous transplant reinforcement in construction or to cure the functional residual instabilities. These implants are either made by the braiding process or by the special flat knitting process and high tenacity polyethylene terephthalate

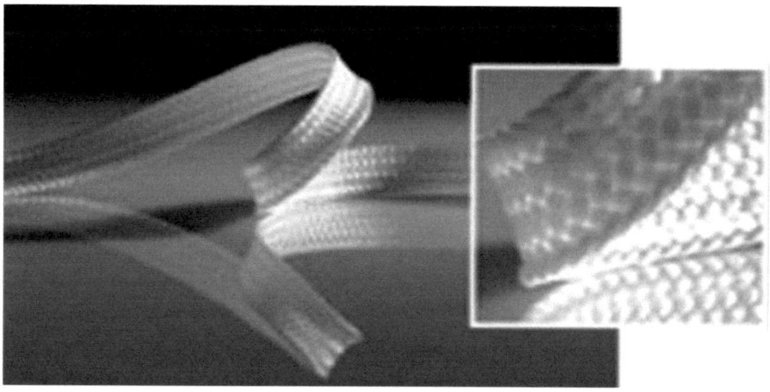

Figure 2.
Woven ligament structure [11].

or high tenacity polypropylene multifilament are used in making the implants for the artificial ligaments (**Figure 2**) [11].

5. Hard tissues

Hard tissue compatible materials must have excellent mechanical properties compatible to hard tissue. Textile structural composites are used for implants. Typical characteristics of polymer related to hard tissue replacement are good processability, chemical stability and biocompatibility. Applications include artificial bone, bone cement and artificial joints. The current practice is to combine bioactive ceramics with polymers or metals to improve interfacial properties. Fiber reinforced composite material may be designed with the required high structure strength and biocompatibility properties needed for these application and are now replacing metal implants for artificial joints and bones.

5.1 Orthopedic implants

Orthopedics is a branch of medicine that deals with disorders with the bones, joints and associated muscles. Orthopedic implants generally serve two purposes, as hard tissue to replace bones and joints, and as fixation plates to stabilize fractured bones. The first orthopedic implants were mainly metal structures. Fracture fixation devices include, spinal fixation devices, fracture plates, wires, pins and screws, adhesives while joint replacement hip, knee, elbow, wrist and finger (**Figure 3**).

The fiber types used for orthopedic implants include polyacetal, polypropylene, and silicone. Composite structures composed of poly (D, L-lactide urethane) and reinforced with polyglycolic acid have excellent physical properties. This sensor principle is designed to allow for a relative strain resolution as small as 10-4–10-5.

5.2 Cardiovascular implants

Due to a steadily growing number of patients and considerable diagnostic and therapeutic advances, vascular diseases are becoming more and more important in general and clinical practices thus the vascular grafts are the need of the hour.

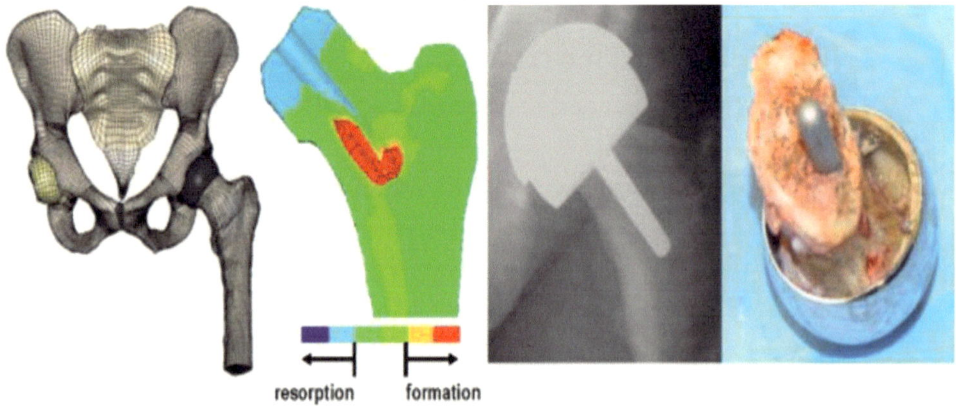

Figure 3.
Hip bone implants [11, 12, 15].

Figure 4.
Knitted structure for a cardiovascular implant [12–15].

Vascular grafts are used in surgery to replace damaged thick arteries or veins. The implantation of synthetic and biological grafts in the circulatory system yield several types of complications ranging from infection to wall rupture. Dilation, suture line failure, structural defects (holes, perforations, rents, and slits), bleeding and infection are some of the main problems caused due to the failure of the grafts. Textile structures are usually the materials used for arterial replacement; however, they do not always meet all the requirements. Gel weave is a true zero-porosity twill woven polyester graft. It is manufactured using an advanced technique of weaving fully texturized polyester on modern looms (**Figure 4**) [12].

The most important aspects of an arterial graft include porosity, compliance, and biodegradability and the design considerations for the graft are selection of the right type of polymer, the type of the yarn, fabric and the crimping. Polyester (e.g. Dacron) or PTFE (e.g. Teflon) and polyurethane are the most commonly. Commercial prostheses contain either single- or two-ply yarns. On one hand these yarns usually have a round cross-section and on the other hand trilobal yarns have been used for the reason it provides the advantage of offering a large surface area making the preclott easier and faster, but they are more prone to fatigue and mechanical damage [13, 14].

6. Non-implantable medical textiles

These are the materials which are used for external applications on the body and may or may not make contact with skin. This includes:

- Wound care
- Plasters
- Orthopedic belts
- Wadding
- Protective eye pads

7. Desired properties for non-implantable medical textiles

Absorbent, wicking performance, non-toxic, breathability, soft, elasticity, non-allergic, ability to be sterilize etc (**Table 3**).

Application	Fiber types	Fabric structure
Absorbent pads	Cotton, viscose, lyocell	Non-woven
Wound contact layer	Alginate fiber, chitosan, silk, lyocell, cotton, viscose	Non-woven, woven knitted
Base layer	Viscose, lyocell, plastic film	Non-woven, woven
Simple non-elastic and non-elastic bandages	Cotton, viscose, lyocell, polyamide fiber, elastomeric fiber yarns	Non-woven, woven
High–support bandages	Cotton, viscose, lyocell, elastomeric fiber yarns	Non-woven, woven knitted
Compression bandages	Cotton, viscose, lyocell, elastomeric fiber yarns	Non-woven, woven knitted
Orthopedic bandages	Cotton, viscose, lyocell, PET, PP, polyurethane form	Woven, knitted
Plaster	Cotton, viscose, lyocell, PET, PP, glass fiber	Non-woven, woven knitted
Gauges	Alginate fiber, chitosan, lyocell, cotton, viscose	Non-woven, woven knitted
Wadding	Viscose, cotton linter, wood pulp	Nonwoven
Lint	Cotton	Woven

Table 3.
Showing application areas and type of fiber used [2–5, 15, 16].

8. Wound dressing

Different types of dressings are available for a variety of medical and surgical applications.

- Functions of wound dressings:
- Protection against infection
- Absorb blood and exudate
- Promote healing
- To keep the wound smooth and pliable
- Medication to the wound

The wound contact layer should prevent adherence of the dressing to the wound and be easily removed without disturbing new tissue growth. Gauge and paraffin coated gauge are the most common dressings used. Most gauges are made from cotton in the form of a loose plain weave. The burns and skin graft sites must have their dressing changed frequency. When the dressing is removed, it is not only painful, but it can also destroy the regenerating tissues. This can delay the healing process because scarring and reopen the wounds for possible bacteria entrance. The paraffin coated gauge which is usually multilayered is a little easier to remove than dry gauge. Gauge may be impregnated with plaster sterilization is required. Finishing agents such as wetting agents and optical whiteners are not added to gauge fabrics because of the possibility of irritation and possible carcinogenic effects [15, 16].

Nonwoven fabrics can be used for the following advantages:

- Better sterilization
- Smooth and lint free (allows for a lesser change for debris to be left in the wound)
- Can be made softer and more absorbent by latex or thermal calendering
- For port operative dressing, sophisticated nonwoven structure is possible. Nonwoven fabrics made of atelocollagen filaments are used as wound dressing for burns.
- Polypropylene fabric/carbonized rayon fabric would transmit liquid to the absorbent material and enable to keep the skin dry.

Wound dressings act as physical barrier for wounds and are found to have some distinguished Properties like fluid control, odor management, and microbial control and wound healing acceleration (**Table 4**).

8.1 Dressing material

8.1.1 Calcium alginate fiber

The basically raw material for the product of this fiber is alginic acid, an emulsion attained from the marine brown algae. It possesses a variety of parcels, including the capability to stabilize thick suspense, to form film layers, and to turn into gels. When

Types	Properties
Passive products	Traditional dressing that provide cover over the wound
Interactive products	Polymeric film to permeable to oxygen but not bacteria
Bioactive products	Dressing that deliver substance active in healing, e.g. alginate, chitogan

Table 4.
Classification of wound dressings [15–18].

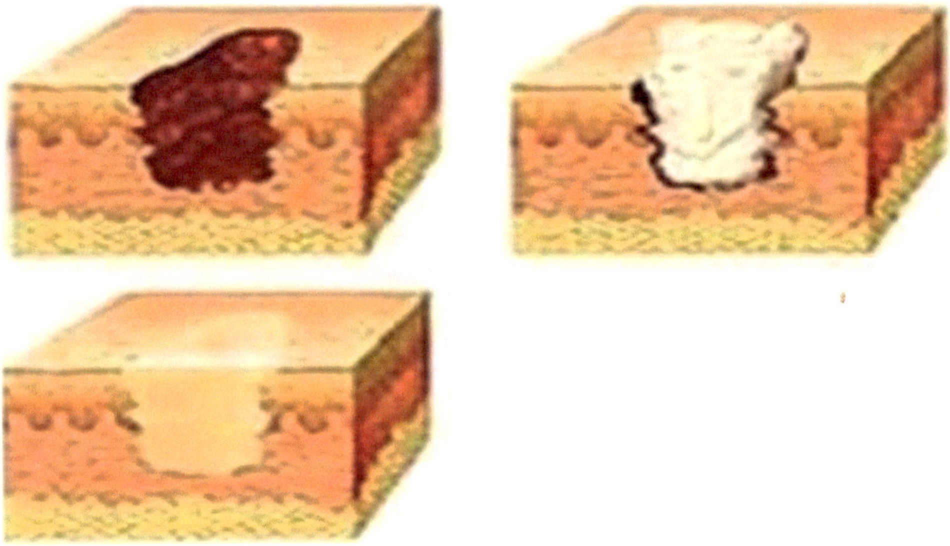

Figure 5.
Sorbalgon wound dressing [4, 15, 16].

the dressing made of this fiber is applied to crack, the rear ion exchange take place and this fiber is placed on the crack in dry state and begin to absorb the exudates.

8.1.2 Sorbalgon

It is a supple, non-woven dressing made from high quality calcium alginate fiber with excellent gel forming properties A Sorbalgon dressing absorbs approximately 10 ml exudates per gram dry weight (**Figure 5**).

8.1.3 Thin film dressing

Thin film has very superior absorbent properties and outer surface thin film give better comfort behavior. This thin layer film has basically working of the easily absorb body fluid and proper safe keep it to the dressing leakage and wound maceration.

8.1.4 Acticoat dressing

Acticoat dressing is give better protection against fungal infection performance as compared to traditional antimicrobial dressing materials. This dressing is better kill rate and more effective fungal species.

9. Bandages

The bandage has generally essential properties should be like breathable, stretchable, non-slip, non-stick to more comfort help during injuries time of human body. Bandages are designed to perform a whole variety of specific functions depending upon the final medical requirement. The basically bandages are used in injuries and wound place to keep it dressing. Such bandages are in form of light-weight knitted fabrics or open-weave woven fabrics, made from either cotton or viscose. Their primary function is to hold the healing wound dressings firmly in place. They themselves do not have healing functions to play [17, 18].

Orthopedic cushion bandages are used under plaster casts and compression bandages to prove padding and prevent discomfort.

Different types of bandages can be classified.

A1: Light weight conforming stretch bandage.

A2: light support bandages.

A3: Compression bandages.

A3 (a): Light compression bandages.

A3 (b): Moderate compression bandages.

A3 (c): High compression bandages.

A3 (d): Extra high performance compression bandages.

9.1 Features of different types of bandages

9.1.1 Compression bandages

It provides necessary support to circumscribe movement and speed up the mending process Compression tapes are used for the treatment and forestallment of deep tone thrombosis, leg ulceration, and swollen modes and are designed to ply a needed quantum of contraction on the leg when applied at a constant pressure. Compression tapes are classified by the quantum of contraction they can play at the ankle and include extra-high, high, moderate, and light contraction and can be either woven and contain cotton and elastomeric yarns or underpinning and weft knitted in both tubular or completely-fashioned forms.

9.1.2 Compression hosiery

- Compression hosiery can be used as an alternative to compression bandaging for the treatment of active ulcers.

- Compression hosiery is classified according to the pressure level applied at the ankle.

- Compression hosiery is made from a number of different fibers including nylon, cotton yarn and elastane.

Figure 6.
Orthopedic bandages [17–20].

9.1.3 Orthopedic bandages

A cloth girth saturated with cataplasm of Paris is dipped into water and also wrapped around the broken branch thereby creating an establishment- fitting yet fluently removed flake in the shape of a tube or cylinder. This type of operation of cataplasm in the form of a broken branch is generally known as an orthopedic cast. The modern plaster fabric is made from spun bonded nonwovens of cotton, viscose, polyester or glass fiber (**Figure 6**) [19].

9.1.4 Pressure garments

Pressure garments play a vital role in the proper healing of wounds and reduce the effects of scaring, but for the garments to perform their job properly, they need to be in good condition. The continuous wearing of pressure garments prevents the thickening, buckling, and nodular formations seen in hypertrophic scars [20].

10. Conclusions

Medical textiles have visible speedy improvement over the previous couple of decades. Nowadays, new biodegradable fibers have enabled the improvement of novel sorts of implants, and contemporary-day fabric machines can produce third-dimensional spacer fabric that supply advanced overall performance over conventional fabric materials.

- These and lots of different advances have made clinical textiles a crucial detail in contemporary-day ailment management, and they are turning into increasingly critical with the growing quantity of aged humans with inside the populations of evolved countries.

- The more significance of medical textiles in human life, healthful residing and development is immense.

- The improvement of latest technology and new gadgets will assist sufferers to conquer the hardships that they used to go through with inside the past.

- There are many extra unknown regions of medical textiles; we must do studies on the ones issues. We must pay extra interest to the manufacturing of healthful and nice clinical fabric materials. In addition to technology, we want to hold a watch at the rate of our products.

- Through this it is going to be viable to supply nice whole and easily to be had contemporary-day medical textiles.

- Textile substances preserve to serve a critical feature with inside the improvement of number clinical and surgical products.

Author details

Ramratan Guru[1*], Anupam Kumar[2], Deepika Grewal[2] and Rohit Kumar[2]

1 Department of Handloom and Textile Technology, Indian Institute of Handloom Technology, Varanasi, U.P., India

2 Department of Textile Engineering, Giani Zail Singh Campus College of Engineering and Technology, Maharaja Ranjit Singh Punjab Technical University, Bathinda, Punjab, India

*Address all correspondence to: ramratan333@gmail.com

References

[1] Kwatra GPS. Textiles for medical applications. Asian Textile Journal. 1992;**2**(4):18-25

[2] Srinivasan J, Kalyana Kumar M. Polypropylene for high performance medical applications. Asian Textile Journal. 2005;**2**(6):73-81

[3] Anand SC, Kennedy JF, Miraftab M, Rajendran S. Medical and Healthcare Textiles. No. 75. New York, Washington, DC: Woodhead Publishing Limited, CRC Press; 2010. pp. 1-555

[4] Desai AA. Biomedical implantable materials sutures. Asian Textile Journal. 2005;**5**(2):54-56

[5] Behrend D, Schmitz KP, Werner C. Resorbable surgical textiles. Asian Textile Journal. 2000;**8**(2):34-35

[6] Hayavadana J, Renuka D. Tissue engineering—A new era in textiles. Asian Textile Journal. 2003;**2**(2):107-110

[7] Gopal Krishnan D, Aravindhan KA. Non-woven for medical textiles. Asian Textile Journal. 2005;**1**(2):36-41

[8] Kalberer K. Medical textiles the textile. Asian Textile Journal. 2000;**8**(6):53-54

[9] Rajendran S, Anand SC. Recent advancements in textile materials for wound management. Asian Textile Journal. 2004;**9**(2):46-51

[10] Koslosky J. Implantable fabrics for orthopedic device applications. Medical Device and Diagnostic Industry. Sep 01, 2007 [Online]

[11] Malik A. Polymers and fibres in disposable medical products. Asian Textile Journal. 2001;**2**(5):36-40

[12] Frank KK, Laurencin CT. 3-D textile scaffolds for tissue engineering. Textile Asia. 2000;**8**(6):27-30

[13] Katzer K. Polyethylene polymers for hygiene market. Asian Textile Journal. 2002;**2**(4):30-36

[14] Bartels VT, editor. Handbook of Medical Textiles. Cambridge: Woodhead Publishing Ltd.; 2011

[15] Qin Y. Medical Textile Materials. Cambridge: Woodhead Publishing Ltd.; 2016

[16] Fisher G. Developments trends in medical textiles. Journal for Asia on Textile and Apparel. 2006;**12**(4):51-68

[17] Berndt E, Geuer M, Wulfhorst B. Dreidimensionale Textilstrukturen zur Herstellungvon technischen Textilien—Stand 2000 (Three-dimensional textile structures for the production of technical textiles). Technische Textilien. 2001;**44**:270-283

[18] Thomas S. Compression bandages. In: Cullum N, Roe BH, editors. Leg Ulcers: Nursing Management. Harrow: Scutari; 1995

[19] Walker IV. Proceedings of Medical Textile Conference. Cambridge: Bolton Institute, U.K. Publishing Co.; August 24 & 25, 1999. pp. 12-19

[20] Kothari VK. Journal of Textile Association. 2006;**67**:181-185

Chapter 2

Healthcare and Hygiene Products Application in Medical Textile

Ramratan Guru, Anupam Kumar and Rohit Kumar

Abstract

Healthcare and hygiene products are usually available over the counter and normally used for hygienic purposes to prevent infection and transmission of diseases, provide hygiene, and enhance care in the hospital ward and operating room. Nowadays it is a scientific research approach to big growing part in medical textiles, in healthcare and hygiene products. The day by day increase in demand of medical textile in different sectors like wipe to operating rooms are more advanced fabrics used with anti-fungal and anti-microbial applications. In this sector, new concepts of low-cost effective techniques are developing day by dayfor both patient and hospital staff to protect them from the effect of virus infection and other bacteria. This paper basically discusses the main role of hygiene and health care sectors application in medical textile.

Keywords: healthcare and hygiene products, design materials, product application and testing

1. Introduction

An important and growing part of the textile industry is the medical and related healthcare and hygiene sectors. The extent of growth is due to constant improvement and innovations in both textile technology and medical procedures. A critical and developing part of the fabric Industry is the clinical and associated healthcare and hygiene sectors. Textile has usually been part of healthcare [1]. The variety of merchandise to be had is sizeable however normally they are used inside the running room theatre or at the health centre ward for the hygiene, care and protection of personnel and patients. The range of programmes variety from the easy cleansing wipe to the superior barrier fabric used for running rooms. Medical textiles constitute systems designed and executed for scientific software [2]. The range of programs is diverse, starting from an unmarried thread suture to the complicated composite systems for bone alternative and from the easy cleansing wipe to superior barrier fabric utilized in running rooms. Textile substances and products, which have been engineered to fulfil precise needs, are appropriate for any scientific and surgical software wherein a mixture of strength, flexibility and from time to time moisture and air permeability is required.

Textile materials and products that have been engineered to meet particular needs are suitable for any medical and surgical applications, where a combination of strength, flexibility and sometimes moisture and air permeability are required [3, 4].

2. Constituent element of medical textile products

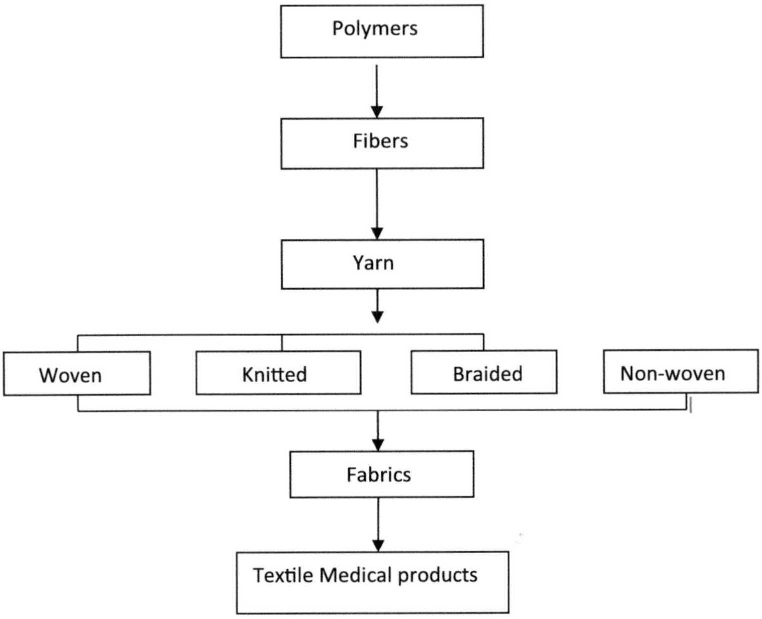

3. Healthcare and hygiene products

Textile has usually been a part of healthcare. The variety of merchandise to be had is sizeable; however, normally they may be used inside the working room theatre or at the health facility ward for the hygiene, care and protection of workforce and patients. The quantity of packages variety from the easy cleansing wipes to the superior barrier fabric used for working rooms [5].

The medical textile fabric merchandise may be prepared into three simple categories

- Patient specific: sponges, sheets, burn sheets etc.
- General patient management: Under-pads, adult diapers and wipes
- Procedure specific: sterilization wrap, surgical gowns, drapes, table covers, face masks, head and shoe covers.

4. Characteristics of materials for medical use

- Nontoxicity
- Non-allergenic response
- The ability to be sterilized
- Elasticity, durability

- Biocompatibility
- Fast and highly absorbent
- Static dissipation
- Antimicrobial

Wide range of staple fibres are used for the hygiene sector and technical application (**Table 1**)

- Bi-component fibres for thermal bonding, used in hygiene articles such as sanitary napkins, baby diapers, etc.
- Special types for hydro-entangled nonwovens are not only employed in the manufacturer of wet and cosmetic wipes but also increasingly in technical applications.
- Fibres made from biopolymers (PLA/Ingeo) include fibres made from renewable plant compounds
- Fill fibres for beds.

Non-woven possesses the following properties due to which they became famous in medical field:

- Various parameters can be controlled easily like porosity, weight of fabric, thickness.
- Nonwovens are easy to sterilise

Used medical application	Type of fibre used	Cloth type
Cover cloths and Surgical drapes	Polyester and polyethylene	Nonwoven or woven
Absorbent layer Outer layer, Incontinence and diaper Cover stock	Polyester, polypropylene Wood fluff and Super-absorbents Polyethylene fibres	Nonwoven
Protective clothing and Clothing uniforms	Cotton, polyester and polypropylene	Woven and Nonwoven
Surgical hosiery garments products	Polyamide, polyester, cotton and elastomeric yarn	Nonwoven and Knitted
Wipes and garments	Viscose rayon	Nonwoven
Masks and caps etc.	Viscose rayon, polyester, viscose and glass	Nonwoven
Pillow covers, sheets and Bedding, Blanket	Cotton, polyester and cotton	Woven and knitted
Surgical gowns garments	Cotton, polyester, viscose rayon and polypropylene	Nonwoven and Woven

Table 1.
Product applications [2–6].

- Various manufacturing technique options according to applications.
- Economical manufacturing process.

5. Textile materials used in operating theatre and emergency rooms

These encompass surgeon's gowns, caps and masks, affected person drapes and cowl cloths of all sizes. The reason for defensive healthcare clothes is to defend healthcare experts from infection from blood and different infectious fluids [6, 7]. Biological defensive clothes are described through the Occupational Safety and Health Administration (OSHA) as follows: 'Personal defensive garb may be taken into consideration suitable most effective if it does now no longer allow blood and different infectious substances to by skip thru to attain an employee's paintings clothes, road clothes, undergarment, pores and skin, eyes, mouth or different mucous membranes below the ordinary situations of use and at some point of time the protection system may be used'. According to this definition, there are fundamental necessities for a defensive fabric garment: it ought to save you infectious substances from passing thru the pores and skin and it ought to closing lengthy enough. Protective clothing inside the clinical subject ought to be affordable, breathable, comfortable, dependable, and effective [8].

6. Design issue

Main issue in design and use of operating room fabric used to be is the protection of the patient from contamination by the environment and by healthcare workers. The principle design features for medical non-woven fabric are barrier properties, strength, sterilization stability, breathability and comfort for garment application.

6.1 Barrier performance

Barrier performance can be partial (resistant) or total (proof) ranging from particulates and bacteria to fluids and viruses. The principal necessities for barrier fabric are that they withstand the penetration of liquids, especially blood and on the equal time be sterile, breathable, bendy and cheaper. Because of those necessities, maximum of the barrier clothes are crafted from non-woven fabric, which can be exceedingly cheaper and may be thrown away after every use, hence lowering the want for re-sterilization. In a few cases, unique breathable movies are being brought to fibres and fabric. In different cases, components are being brought without delay into polymers getting used to making the fibres. Theatre drapes are meant to shape a barrier in opposition to contamination each to and more good from the patient. Strength requirements vary with end-use application. For surgical drapes, stiffness is very critical because barrier performance may be affected by comorbidity to patient or equipment. Good abrasion resistance is necessary for the safety of barrier administration. The consumer product safety commissions (CPSC) require 3.5sec burn time on CS-191-53 for gowns, head covering and surgical mask [9–11].

6.2 Sterilization stability

Many hospitals have delivered peroxide plasma systems, inclusive of STERRAD, to their steam autoclaves and ethylene oxide chambers inside the Central Supply Room.

In designing fabric for sterilization, it is far critical to apprehend the effect of sterilization tactics on material overall performance features. Steam autoclaves typically function at 132°C. Fabrics containing cellulose are not typically advocated for the plasma gadgets as those fabrics hold residual peroxide.

6.3 Comfort and breathability

The consolation and breathability components are commonly taken into consideration as opposing the barrier performance. For sterilization wrap, the difficulty is that the barrier should save you dirt and micro-organisms from penetrating a sterilized bundle in the course of garage and transportation. At an equal time, it should be porous sufficient for the sterilant to penetrate the wrapped bundle and absolutely sterilize the content material of the surgical set.

6.4 Linting

For gowns, linting is not wanted because particles from gowns or drapes may complicate the wound healing process. In general, it is accepted that particles above 50μm are readily visible to the unaided eye.

6.5 Antimicrobial textiles

Treated fabric articles can encompass clinical textiles consisting of pads, face masks, surgical gowns, ambulance blankets, stretchers, and clear out substances and diapers [12].

6.6 Antimicrobial fibres

High overall performance fibres had been evolved which save you risky microorganism from buildup and could discover programs with inside the fields of private hygiene wherein buildup of risky microorganism may be risky to health: the fibre essentially includes a mixture of antimicrobial compounds, primarily based totally on steel salts which in the end controls microorganism and fungi. The compounds are embedded inside the matrix of fibres which renders it impervious to washing and wear [13].

7. Product application

7.1 Surgical gowns

Surgical gowns are worn during medical procedures, to prevent contamination by splattering of body fluids such as blood, respiratory secretions, vomit or feces during medical procedures.

Surgical gowns are made of fluid-resistant materials to reduce the transfer of body fluids (**Figure 1**). Isolation gowns are usually intended to protect the wearer from the transfer of micro-organisms and only small amount of body fluids [14].

7.2 Surgical masks

They should have high level of air permeability, high filter capacity, and should be light weight and non-allergic.

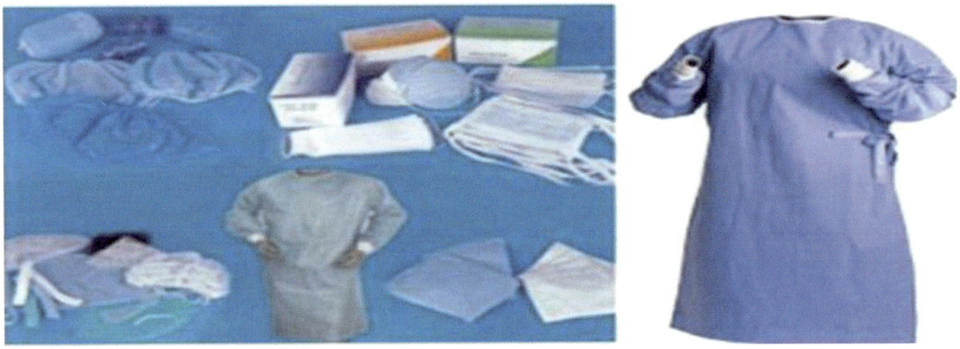

Figure 1.
Surgical gowns, healthcare and hygiene products [9–11].

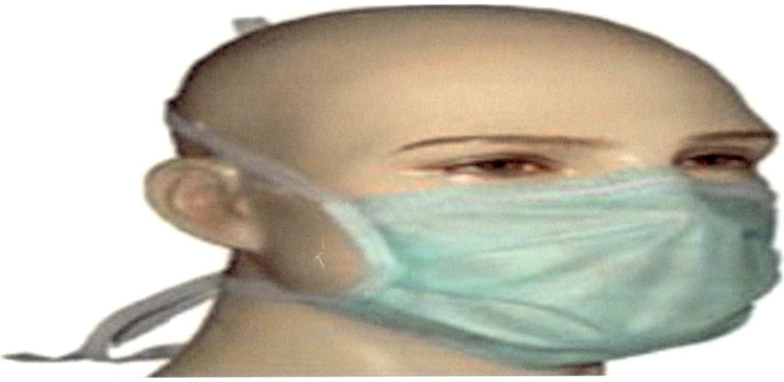

Figure 2.
Surgical masks products [14, 15].

Materials: consist of a very fine middle layer of extra fine glass fibres or synthetic micro fibres covered on both sides by either an acrylic bonded parallel-laid or wet-laid non-woven (**Figure 2**).

7.3 Surgical drapes and cover cloths

These are used to cover patients or working areas around patients.

Material: loop raised warp-knitted polyester fabric laminated with PTFE films for air permeability, comfort and resistance to microbiological contaminants [15].

7.4 Surgical hosiery

Surgical hosiery with graduated compression traits is used for wide variety of purposes, starting from a mild help for the limb to the remedy of venous disorders. Knee and elbow caps, which might be commonly fashioned throughout knitting on round machines and might additionally comprise elastomeric threads are worn for help and compression throughout bodily energetic sports activities or for protection.

7.5 Hospital ward textiles

Such as bedding garb, bed covers, incontinence merchandise are used for care and hygiene of patients. The conventional Woollen blankets had been changed with cotton leno woven blankets to lessen the threat of pass contamination and are crafted from smooth spun two-fold yarns which own suited thermal qualities. In isolation wards and in-depth care units, disposable defensive garb is worn to reduce pass contamination and are made from composite of tissue strengthened with a PET or polypropylene spunlaid web [16].

7.6 Cleaning products

These include gauze for floors, dry dusting systems; hard surface disinfectant wipes high absorbency cloth, window cloth, electrostatic disposable dusters, cleaning mop, etc.

8. Absorbent hygiene products

8.1 Modern breathable disposable feminine products

Figure 3 classify of three layers:

- Inner pinnacle layer: fabricated from a mix of hydrophobic low-density fibre and is liquid and water permeable.
- Core layer: full of wooden pulp and different absorbent fabric is especially absorbent.
- Third layer encompasses multi-layer barrier, this is water vapour permeable, however persistent to liquid water.

8.2 Modern incontinence product

Modern incontinence product also consists of three layers.

Cover stock that is permeable and diffuses liquid laterally. Highly absorbent core and barrier polyethylene or PVC films that help patient cloths or bedding to keep dry.

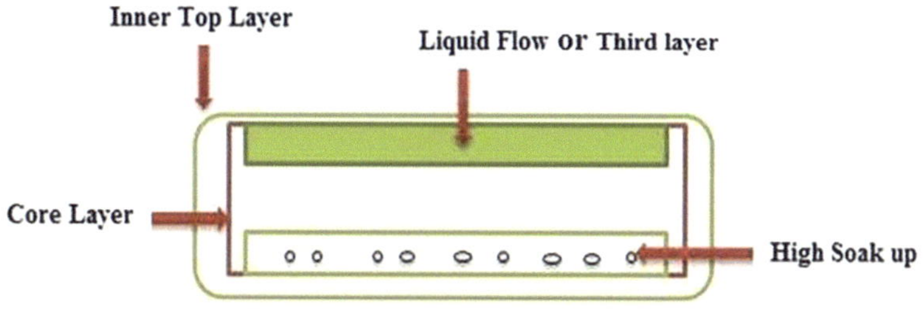

Figure 3.
Modern breathable disposable feminine products [16, 17].

9. Testing of healthcare garments

Laboratory exams consist of water repellency, launder ability, burst electricity and tear electricity. The layout of barrier fabric is pushed with the aid of using the priority over HIV. Therefore, for those fabric check techniques that might help with inside the Characterisation of merchandise as blood-resistant, blood-evidence or viral evidence. These techniques were installed as ASTM 1670-95 and 1671-97.

The call for wettability approach of measuring the absorbency traits of fabric was defined with the aid of using Lichstein. This method measures each capability and absorption price concurrently at zero hydrostatic head. It is relevant to distinctive absorbents, wicking fluids and more than one ply system with the absorbent at any attitude to the fluid and below distinctive pressure [17].

10. Advanced medical textiles

Bio-purposeful substances are starting up new opportunities for the medical fabric sector. Here energetic materials are included in the fibre with the aid of using chemical change or implemented onto the fibre floor at some point of the spinning process. These components are transferred to the pores and skin with the aid of using frame moisture and frame warmth with stepped forward bioclimatic and hygienic homes such as.

- Protection to the pores and skin from liquids, debris and bacteria.
- Providing a powerful barrier towards germs, fungi and danger of infection.
- Thermo regularity characteristic.
- Ease of laundering, sterilization and anti-static behaviour
- Low stage of fabric chemical compounds and dyes with excessive mechanical stability.

New fabric is evolved to face up to bacteria, mildew, stain and odour for healthcare applications. For example, anti-allergen completing retailers are used on material to offer themselves remedy to sufferers from bronchial allergies and allergic reactions as a result of dirt mites.

Active substances can also be made available to the skin as an aqueous solution by micro-encapsulation or by their insertion into water-absorbing network polymers, which are affixed to the fibre. Advanced processes also offer the potential for the development of bioactive, drug-delivering textiles and the controlled treatment of diseases.

11. The functional requirement of bedding material for elderly patients

Ideal bedding materials attribute the following

- Convenient during changing and to wash
- Breathable

- Absorbent
- Durable
- Odour free
- Now allergic
- Hygienic
- Easy to care for and store
- Prevent liquid and soil spread around
- Colour and pattern of the fabric (not significant for elderly)
- Provide comfort support
- Keep patient clean after being soiled
- Smooth to the touch
- Keep patient warm after being wetted
- Soft to touch dry after being wetted

12. Conclusions

The application of textile in high performance and specialized fields are increasing day by day. There will be an increasing role for medical textiles in future.

- Textile substances hold to serve a critical characteristic with inside the improvement of number clinical and surgical products.
- The advent of recent substances, the development in manufacturing strategies and fibre properties, and the usage of greater correct and complete trying out have all had a giant impact on advancing fibres and fabric for clinical applications.
- Advances in nonwovens have ended in a brand new breed of clinical textiles. Advanced composite material containing a combination of fibres and fabrics has been developed for applications where biocompatibility and strength are required.
- As medical procedures continue to develop, the demand for textile material is bound to grow.

Author details

Ramratan Guru[1*], Anupam Kumar[2] and Rohit Kumar[2]

1 Department of Handloom and Textile Technology, Indian Institute of Handloom Technology, Varanasi, Uttar Pradesh, India

2 Department of Textile Engineering, Giani Zail Singh Campus College of Engineering and Technology, Maharaja Ranjit Singh Punjab Technical University, Bathinda, Punjab, India

*Address all correspondence to: ramratan333@gmail.com

References

[1] Scarlet R, Deliu R, RozemarieManea L. Implantable medical textiles: characterization and applications. In: 7th International Conference-TEXSCI. Liberec, Czech Republic; 2010

[2] Kwatra GPS. Textile for medical applications. The Indian Textile Journal. 1992:18-23

[3] Samash K. Medical textiles: A healthy prognosis. Textile Month. 1989:15-17

[4] Fisher G. Developments trends in medical textiles. Journal for Asia on Textile and Apparel. 2006;**12**(4):51-69

[5] Walker IV. Proceedings of Medical Textile Conference. Cambridge: Bolton Institute, U.K. Publishing Co.; 1999. pp. 12-19

[6] Gupta S. Innovations Fuel the Use of Nonwoven-Based Medical Textiles. Journal for Asia on Textile and Apparel; 2011;**2**(4):42-48

[7] Gupta BS. Medical textile structures: An overview. Medical Plastics and Biomaterials Magazine. 1998;**2**(1):36-48

[8] Anand SC. University of Bolton, UK. Healthcare and Hygiene Products. Medical Textile and Biomaterial for Healthcare. Cambridge, England: Woodhead Publishing Ltd; 2005. pp. 77-79

[9] Rigby AJ. Textile materials in medicine and surgery. Textile Horizons. 1994:42-46

[10] Adanur S. Wellington Sears Handbook of Industrial Textiles. USA: Technomic Publishing Company, Inc, Auburn University; pp. 348-352

[11] Prescott BA. Proceedings, Fibres to Finished fabrics. The Textile Institute. London: Northwestern Oklahoma State University; 1998. pp. 213-216

[12] Desai AA. Special textiles used for manufacturing healthcare and hygiene products. Textile Magazine. 2003;**73-76**

[13] Walker IV. Textiles and sterilization assurance. Proceedings, Medical Textiles International Conference, Stuggart; 1989. pp. 14-22

[14] Fisher G. Medical and hygiene textiles continuing in good health. Technical Textiles International. 2002;**2**(4):10-15

[15] Anon. Silver oxide antimicrobial textiles. Medical Textiles, International Newsletters. 2003;**2-6**

[16] Holla U. Medical textile industry poised for a healthy growth. Express Textile. 2000;**2**(6):42-52

[17] Smith BM. Steadfast. A leader in Canadian medical textiles. Canadian Textile Journal. 1999;**14-15**

Chapter 3

Bacterial Cellulose: Biosynthesis and Applications

Ahmed Amr and Hassan Ibrahim

Abstract

Bacterial cellulose (BC) or microbial cellulose (MC) was considered a bioactive material characterized by high absorbed water, high crystalline, high tensile strength, and biodegradability. However, bacterial cellulose has wide applications, such as biomedical, textile, paper industries, food, drug release, and cosmetic applications. So the microbial cellulose production from Acetobacter *xylinum* from different wastes such as carbon and nitrogen sources, for example, pineapple peel juice, sugar cane juice, dry olive mill residue, waste beer yeast, and wheat thin stillage, are characterized by FTIR, XRD, SEM, and TEM. The product yield of bacterial cellulose is affected by different factors such as the concentration of sugar in carbon source, temperature and time of incubator of the strain, and pH of media. So, it must be studied with the enzymatic pathway procedure.

Keywords: bacterial cellulose, microbial cellulose, synthesis, biomedical, applications

1. Introduction

Natural polymer cellulose is synthesized by plants, as well as fungi, algae, and bacteria, while the structure of a bacterial cell has rarely a ratio to cellulose. The cell wall of plants and seed shells, wood, contains the main component is cellulose macromolecules that are constructed from an unbranched D-glucose chain. However, the glucose units are connected with each other by a 1,4-β-glycosidic linkage in a linear form. The length of cellulose polymer chains depends on the nature of the producer and accordingly differs among themselves [1]. The main sources of produced cellulose are of four different types: The first is produced from plants, the second method is the preparation of cellulose from various microorganisms, fungi, and algae, and the remaining two are less common: the first is the synthesis from enzyme *in vitro*, beginning with cellobiose fluoride, and the second is synthesis from glucose chemically by opening the ring polymerization of benzylated and pivaloylated derivatives. The cellulose from plants and bacteria is the same molecular form of $(C_6H_{10}O_5)_n$; however, the chemical and physical properties are different (**Figure 1**). Plant cellulose is different from the bacterial cellulose by its low crystallinity, low water absorbing capacity, and ultrathin structure [3, 4].

There are some components that are often found in plant-derived celluloses, including lignin, hemicelluloses, and pectin, but these components are absent from bacterial cellulose (BC), which is highly purified relative to plant cellulose and hence

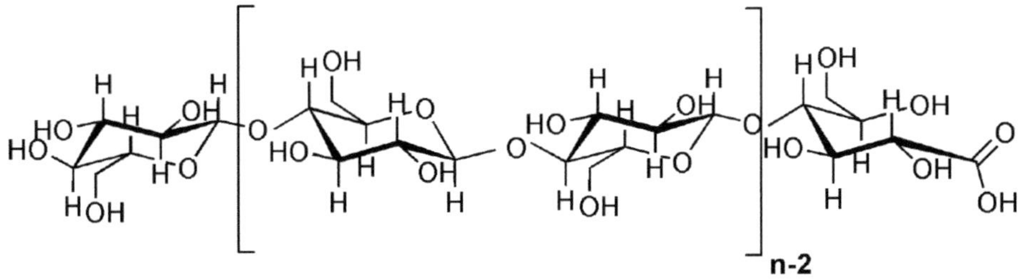

Figure 1.
The 1,4-β glycoside chain of cellulose [2].

requires a low-energy procedure [4]; usually, we use more chemicals in purification of cellulose produced from plants, which is low energy consumption process [5]. Regardless the plant cellulose is constructed from polysaccharide based on glucose units, this is the basic material of plant cell walls, which is utilized as crude materials in paper, pulp, fabric, and textile industry (as 10 hydrogen-bonded chains, each with 500 to 14,000 1,4- β-linked glucoside molecules) [6]. In a cellulose chain, 1, 4-β glycoside bonds are present in order to link the D-glucose pyranose units as a linear polysaccharide **Figure 1**. The length of chain is approximately 0.3 nm wide [7].

Usually, natural polymer must be biodegradable, renewable, and bioactive compound, which is characterized by a high modulus, high mechanical strength, and low density; therefore, during the processing, it is more difficult to damage, and for processing equipment we have some requirements, and cheap raw material [8]. The plant cell wall is used to isolate the cellulose. So, there are different sources of cellulose, including wood, pulp, and cotton. After the long fibers are removed from cotton seed, the short fibers remain. Also, it can produce cellulose from the fibers of plants, the plant that produce the cellulose like bagasse (sugar cane stalks), soybean hulls, oat hulls, rice hulls, corn cobs, wheat straw, bamboo, sugar beet pulp, yarn of jute, ramie, and flax [9].

The natural cellulose polymer has a number of glucose molecules about 10,000 [10]. The cellulose chain includes inter- and intramolecular hydrogen bonds, where the free rotation of ring is hindered and the hydrogen bonds of cellulose chain caused the stiffening of chain, and is insoluble in common solvents. In fact, cellulose is a natural polymer. It has hydrophilic properties and contains two hydroxyl groups; one is secondary and the other hydroxyl is primary. However, due to water adsorption of cellulose, it has these hydroxyl groups in chain [11].

Bacterial cellulose (BC) or microbial cellulose (MC) was considered a bioactive material, which is more characterized by high crystalline, high-absorbed water holding, high tensile strength, and biodegradability. Due to the better aforesaid characteristics of bacterial cellulose, it is supported for many human applications, such as in textile and paper industries, food, drug release, medical fields, and cosmetics.

Compared to the high cost of the commercial culture media, bacterial cellulose (BC) production is more expensive. So, researchers study to change different formulations in the food source of strains such as yeast extract and glucose, to lower the cost of food source, and hence, the cost of the production of BC is reduced.

The high purification of cellulose can be produced from several bacteria. The *Acetobacter xylinum* species is used to produce bacterial cellulose, its nomenclature, the genus *Gluconacetobacter* as *Gluconacetobacter xylinus* [12]. The characterization of bacterial cellulose such as degree of polymerization (Dp) between 2000 and 6000 [13]. The cross-sectional diameter is between 2 and 4 nm [14], and crystallinity is up

Property	Plant cellulose	Bacterial cellulose	References
Fiber width, mm	$1.4–4x10^{-2}$	70–80	[15]
Crystallinity, %	56–65	65–79	[16]
DP	13,000–14,000	2000–6000	[17]
Young's modulus. GPa	Cotton 5.5–13	BC sheet 15–30	[15]
Water content %	Jute-27	BC fiber 120	
	Flax-28	BC crystal 138	
	60	98.5	

Table 1.
Comparison between BC and plant cellulose.

1.	High purity
2.	High degree of crystallinity
3.	Sheet density from 300 to 900 kg/m^3
4.	High tensile strength
5.	High absorbency
6.	High water binding capacity
7.	High elasticity, resilience, and durability
8.	Nontoxicity
9.	Metabolic inertness
10.	Biocompatibility
11.	Susceptibility
12.	Good shape retention
13.	Easy tailoring, physiochemical properties

Table 2.
Characteristics of BC [16].

to 60%. It has excellent shape and strength retention. **Table 1** shows the comparison between plant cellulose and bacterial cellulose BC (**Table 2**) [16].

2. Synthesis of bacterial cellulose from *Gluconacetobacter swingsii* sp.

2.1 Sugar cane juice and pineapple peel juice were used as food culture source

There are a few animals and some number of bacteria, such as *Gluconacetobacter* (named Acetobacter) [18, 19]. This is a strictly aerobic and gram-negative bacterium; at certain conditions, such as incubator (25 to 30°C and pH from 3 to 7) [13, 20], the bacterial cellulose production uses carbon sources such as glucose, fructose, sucrose, mannitol [21, 22]). The bacteria take three processes to synthesize bacterial cellulose. In the first process, the polymerization of glucose molecules forms the cellulose chains, where the molecules are linked by β -1,4- glucosidic linkages each. Nearly, 1.5-nm-wide protofibril consists of 10–15 equal parallel chains. Then, in the second step, 2–4-nm-wide protofibrils have been collected to form microfibrils, and, in step three, the microfibril groups are collected into a 20–100-nm-wide ribbon. After the former steps, the pellicle of bacterial cellulose [13, 23] produces a matrix of interwoven ribbons.

Hestrin and Schramm's medium is used for producing bacterial cellulose [24]. The cellulose microfibrils are synthesized in different media.

From homemade vinegar, culture can isolate *Gluconacetobacter* strain, identified by 16S rRNA method [25], as *Gluconacetobacter swingsii sp.* [26], sucrose, 0.23%, w/v, sugar cane juice (0.008%, w/v, fructose, 8.57% w/v, glucose, 0.066%, w/v, total nitrogen), pineapple peel juice (2.4%, w/v, fructose,2.14%, w/v, glucose, 2.10%, w/v, total nitrogen, sucrose, 0.31%, w/v) were used as culture media for producing bacterial cellulose, and Hestrin-Schramm (HS) medium (0.5%, w/v, peptone, 2%,

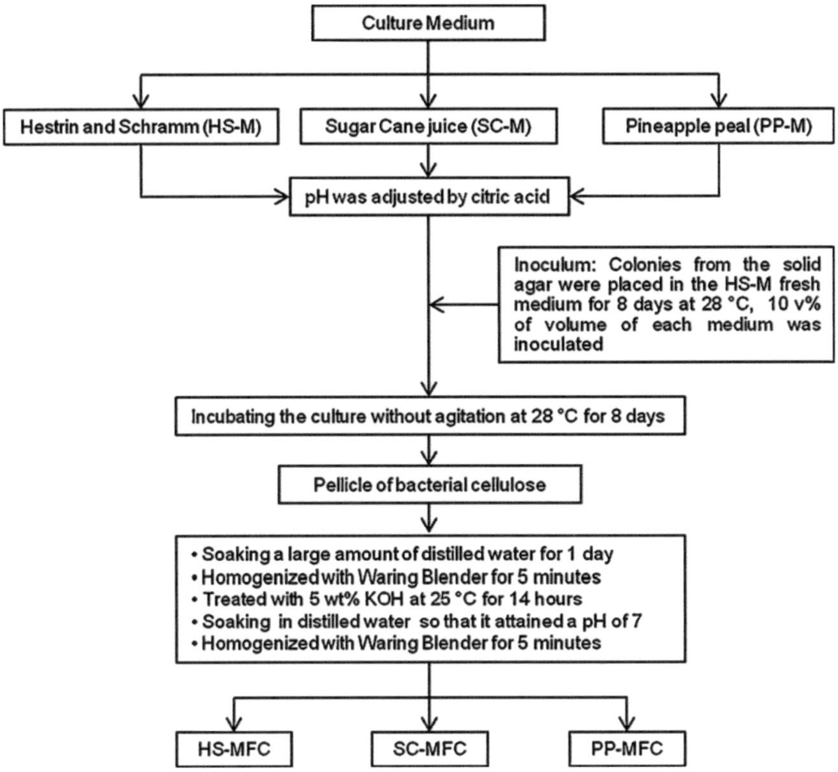

Figure 2.
Steps of BC production using different culture media.

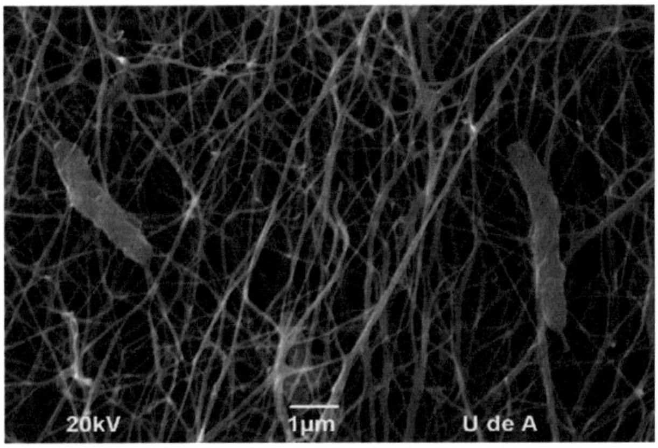

Figure 3.
SEM picture of cellulose ribbons with attached homemade vinegar pellicle and bacteria.

w/v, glucose, 0.5%, w/v yeast extract, 0.27%, w/v, Na_2HPO_4). In the HS medium, the nitrogen source, peptone, and yeast extract are very important [18].

The cellulose obtained from different media is summarized in **Figure 2**. Three culture media will be written as SC-MFC, PP-MFC, and HS-MFC; consequently, after 13 days, 28°C and pH at 7 give bacterial cellulose [18], where the characterized of the bacterial cellulose by SEM and TEM gives this image in **Figures 3** and **4**. The picture of scanning electron microscopy (SEM) has the rode shape of the surface of pellicle formed. TEM picture shows negatively stained specimens of typically 20–70-nm-wide ribbons.

The high yield of bacterial cellulose production using Hestrin and Schramm's medium has similar properties to that produced using pineapple peel juice. The result amount of bacterial cellulose using pineapple peel juice is (2.8 g/L), which is higher than produced by Hestrin and Schramm's medium (2.1 g/L) [18]. Thus, it can be produced BC, with low-cost sources in order to increase its production of bacterial cellulose.

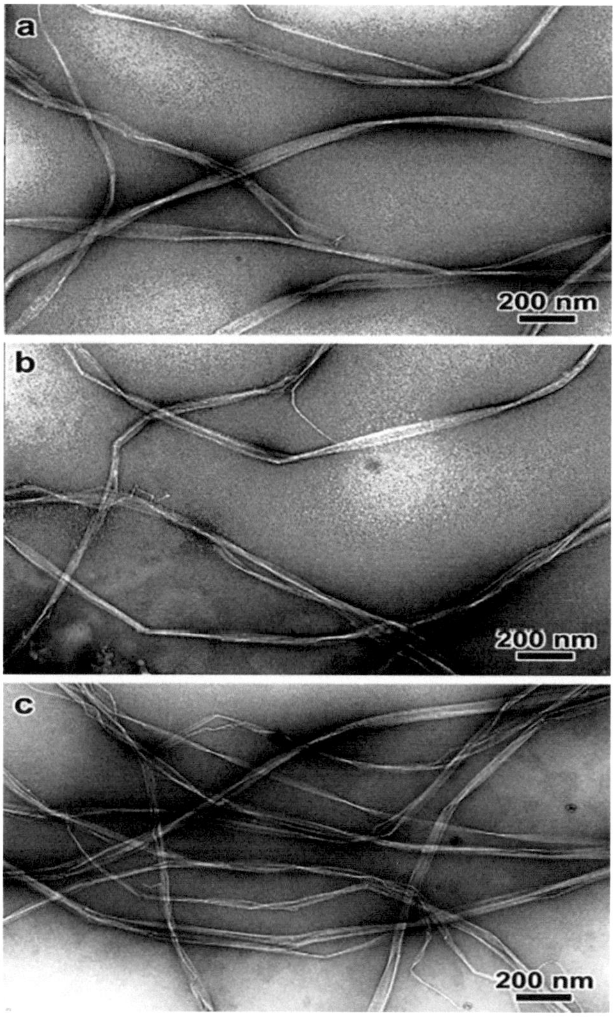

Figure 4.
TEM image of biosynthesized bacterial cellulose ribbons: (a) HS-MFC, (b) SC-MFC, and (c) PP-MFC, prepared by negatively stained specimens of typically 20–70-nm-wide ribbons.

2.2 Characterization of bacterial cellulose produced by *Gluconacetobacter swingsii* sp.

We can see, in **Figure 3**, to show the pellicle surface was formed by *Gluconacetobacter swingsii sp*. from homemade vinegar in rod shape, which is examined by SEM. We have observed the three-dimensional cellulose microfibrils [27], arising from the cell surface and forming bundles. **Figure 4** shows typical 20–70-nm-wide ribbons; this is examined by TEM images recorded from negatively stained species, and a thickness between 6 nm and 8 nm was estimated. Therefore, the cellulose microfibrils consist of 3–11 ribbons [28].

The bacterial cellulose from pineapple peel juice and sugar cane juice with low-cost resources is increasing production to a larger scale [18]. Compared with Hestrin and Schramm's medium, it gives low yield of bacterial cellulose with similar characteristics of these results.

3. *Gluconacetobacter sacchari* using dry olive mill residue produces bacterial cellulose

Wastes from many industries can be used successfully to produce BC, and Japanese pear and grape [29, 30], sugarcane molasses, Konjac powder [31], corn steep liquor [32], many fruit juices, such as apple, orange, pineapple, and beet molasses [33] as well as coconut water [34], are investigated. Due to high cost in the production of BC because it uses quite expensive culture media, the aim of this work is the utilization of residues from the dry olive mill residue production industry as food for *G. sacchari* to evaluate the possible presence of carbon source for the production of BC. On the other hand, it was using conventional HS culture medium to produce BC at nearly 2.5 g/l [29]. So, this study hydrolyzed DOR by acid, after hydrolysis by dilute acid, in order to give compound containing sugars and carbon for food source of BC production.

3.1 Producing bacterial cellulose from extract dry olive mill residue (DOR)

To prepare source of sugar-rich aqueous extracts from dry olive mill residue (DOR40 and DOR100, respectively) to produce BC, it has two water extraction, one at 40°C and second at 100°C. **Figure 5** shows that when lower amount of BC is produced in case of DOR40 and DOR100, the yield of BC is equal to 0.81 and 0.85 g/l compared with conventional HS culture medium (2.5 g/L). This study shows a decreased amount of BC resulting in case of the conventional HS culture medium with a relative ratio from 32 to 34%.

In the step, during hydrolysis of dry olive mill residue (DOR 100H) no BC results, because during the hydrolysis process, the monosaccharide is produced with some of the organic compounds such as furfural, and the phenolic compounds were the results from the degradation of sugar, which could have damaged the metabolism of *G. sacchari*. So the BC is produced during the two aqueous extracts DOR40 and DOR100 but in the aqueous extract at 40°C, low energy is consumed [29].

3.2 Supplemented with N and P sources to produce BC from DOR residues

In order to improve the production of BC, this work can be used as the source of nitrogen and phosphate as supplements with the extract of aqueous DOR 40. These sources were potassium dihydrogen phosphate (KH_2PO_4) and ammonium sulfate $(NH_4)_2SO_4$, respectively. There is slight decrease of BC at high concentration of ammonium sulfate and increase of BC yield at low concentration of ammonium sulfate $(NH_4)_2$ SO_4 at 1 g/l of salt. But, after the addition of 1 g/l of KH_2PO_4 there is slight decrease in the production of BC. This study indicates use of ammonium sulfate and potassium dihydrogen phosphate as a source for increased yield of BC. So, these data indicate that phosphorus and nitrogen sources could also play important roles for the production of BC. But this result in lower conventional HS media.

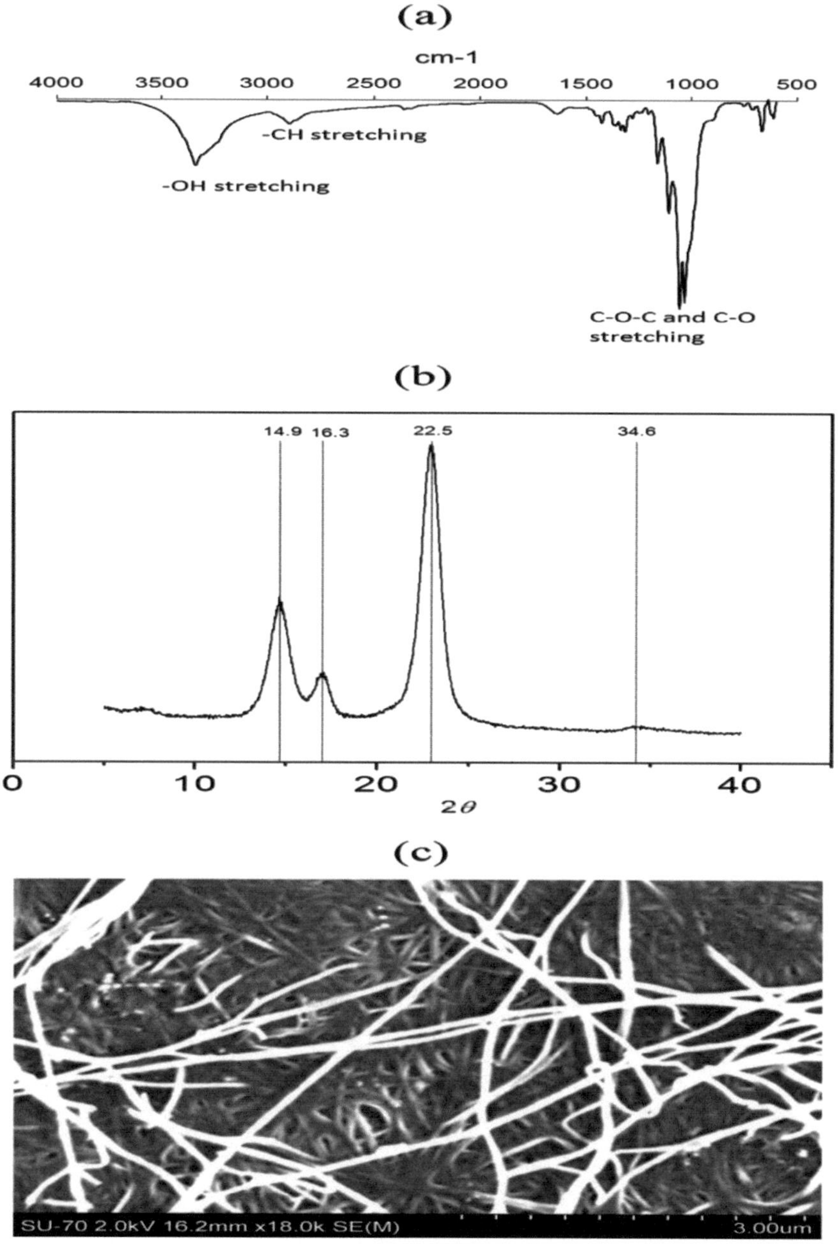

Figure 5.
(a) FTIR spectrum, (b) X-ray diffractogram, and SEM image of the BC membrane obtained with DOR40.

3.3 Characterization of bacterial cellulose (BC)

Production BC is studied when dry olive mill residue water extract is used as nutrients and source of carbon in the presence of nitrogen and phosphate salts. Crystallinity, chemical structure, and morphology of BC can be characterized by XRD, FTIR, and SEM, respectively. **Figure 5a** shows the absorption peak of FTIR for BC appears strong at 2880, 3300, and 1100 cm^{-1}, indicating the vibrations of the CH, hydroxyl group (OH), and C-O-C functional of (BC) and **Figure 5b** indicates the crystallinity, where the presence of the diffraction peaks is at 2 Θ, 14.9, 16.3, 22.5, and 34.6 and crystallinity of these samples is at 80%. The image of SEM is in **Figure 5c** showing homogeneous nano- and microfibrils of cellulose in the tridimensional network.

4. *Gluconacetobacter hansenii* CGMCC 3917 using only waste beer yeast as a nutrient source for biosynthesis of bacterial cellulose

However, using industrial materials waste would not only reduce environmental pollution to a high degree but also improve the production of cellulose by microorganisms. In general, waste beer yeast (WBY) is composed of 23–28% carbohydrate, 48–55% protein, 2% vitamin B, 6–8% RNA, and 1% glutathione; also, it has some elements such as K, Ca, Fe, P, and Mg [35, 36]. Microorganisms could use its sufficient food supply to produce natural green materials. Additionally, large-molecular-weight polymers made of proteins and carbohydrates can be found in the cell walls. So, it is more difficult to utilize it directly as a food source for microorganisms [37].

In this study, it can be produced by reducing sugar yield from waste beer yeast (WBY) by two-process pre-treatments. The first was treated by four methods, including a) high-speed homogenizer, b) 0.1 M NaOH treatment, c) microwave treatment, and d) ultrasonication. The second step is using mild acid condition (pH 2) for hydrolysis at 121°C for 20 min and after this pre-treatment must be evaluated for reducing sugar. While this was modified, the hydrolyzed WBY was directly used as food media culture for *G. hansenii* CGMCC3917 to produce BC. This bacterial cellulose (BC) can be evaluated by 1) water-holding capacity (WHC), 2) water absorption rate, and 3) water release rate (WRR) (WAR) estimated and its microstructure was evaluated using scanning electron microscopy (SEM).

4.1 Production of bacterial cellulose by *G. hansenii* CGMCC 3917 strain

In order to isolate this strain from homemade vinegar, it was recorded as CGMCC3917 at China General Microbiological Culture Collection, Beijing, China, and it was kept on glucose agar slants, including 2% glucose (w/v), 1.5% ethanol (v/v), 0.1% K_2HPO_4 (w/v), 1.7% agar (w/v), 0.5% yeast extract (w/v), and 1.5% $MgSO_4.7H_2O$ (w/v). It was put in a refrigerator at 4°C for every 2 months for inoculum development sub-cultured or deposits at 80°C, and this process must be occurring instead of agar for long-period storage using 20% (v/v) glycerol [38].

4.1.1 Pre-treatments and hydrolysis of waste beer yeast (WBY)

This pre-treatment occurred by taking different concentrations of dry waste beer yeast between 5%, 10%, 15%, and 20% (w/v), respectively, in a 250-ml round-bottom flask by adding 100 ml of distilled water to it, by using unmodified WBY mixed liquor [35]. Four processes were explained to modify WBY mixed liquor. The modification method was as follows:

Pre-treatment 1: In this process of modification, the different amount of WBY is taken, and then, a certain solution from 0.1 M NaOH is taken at 50°C for different interval periods of 6, 12, 18, 24, 30, and 36 h, respectively.

Pre-treatment 2: mixed liquor for 5, 10, 15, and 20 min, respectively.

A certain different weight from WBY is modified by homogenizer at 15,000 rpm (XHF-D, Ningbo Xingzhi Biotechnology Co., Ltd., Zhejiang, China) Pre-treatment 3: by ultrasonication modification for 10, 20, 30, 40, 50, and 60 min respectively, while the ultrasonicator has power of 500 W (YQ-1003A, Ningbo Power Ultrasonic Equipment Co., Ltd., Zhejiang, China). Pre-treatment 4: modified mixed liquor with different concentrations of WBY in microwaves at microwave power of 600 W (Galanz P70D20P-TF, China) for 5, 10, 15, and 20 min, respectively, after each pre-treatment, hydrolysis of mixed liquor samples is carried out with different concentrations of WBY. Pre-treatment is carried out under dilute acid at 121°C for 20 min (pH 2.0).

4.1.2 BC production using WBY hydrolysates with different concentrations

For the production of BC, the highest reducing sugar yield was selected after modifying the waste beer yeast (WBY). After pre-treatment, at 121°C for 20 min, in the presence mild acid condition at (pH 2), WBY was hydrolyzed for 15 min using centrifugation at 4000 g to remove precipitate and the amount of solution was collected and added.

(50%, w/v) glucose solution was prepared using sterilized water in a glass vessel (500 mL) from its initial reducing sugar of 4.38% (w/v) containing 100 mL of WBY hydrolysates at different concentrations of between 1%, 3%, 5%, and 7% (w/v), respectively and after adjusting its pH to 5 using 2 M NaOH, the prepared seed inoculums with (9%, v/v) were stored to cultivate at 30°C for 14 days. The production of bacterial cellulose was from WBY hydrolysates as carbon and food sources without any extra nutrient added. They were directly supplied to G. hansenii CGMCC 3917 to produce high yield of bacterial cellulose [35].

We notice that the samples do not centrifuge after these pre-treatments have a high sugar concentration. So, they see that inhibition of the BC production decreases the supply of oxygen by the liquid medium in case of uncentrifugation. While, in case of using the centrifuge for samples, It reduces sugar by adding water to the supernatant; this is referred to as diluting and gives a better yield for BC production. In contrast, pre-treatments 2 and 4 and unmodified WBY give a lower BC result compared to those not using the centrifuge and centrifuged WBY. Likely, the *G. hansenii* CGMCC 3917 strain type was damaged and does not give bacterial cellulose, because of decrease in sugar concentration present in the centrifuged samples. Additionally, pre-treatment method 3: the WBY is treated by ultrasonication to produce the highest yield of BC (3.89 g/L). Further, the amount of BC from pre-treatment 1 by 0.1 M NaOH is 2.33 g/L [35].

It is clear that the amount of BC in pre-treatment processes 2,4, and 5 was decrease in processes 1 and 3.

4.1.3 Effect of culture time on the production of BC

The concentration of reducing sugar affected BC production. So the ultrasonication method is used to give the highest reducing sugar to improve the BC production, and we must select this to investigate the optimal sugar concentration. The BC weight utilizing WBY hydrolysates in the presence of 1% sugar was the lowest yield and it reaches a value 2.18 g/L on day 7. When, using the sugar of 3%, the BC weight is 7.02 g/L on day 10 [35].

5. Biosynthesis of bacterial cellulose by *Gluconacetobacter sucrofermentans* B-11267 using wheat thin stillage

The thin stillage of rice (R-TS), the wine distillery of rice, was giving wastewater that has organic acids. Recently, researchers are working to use stillage in order to obtain a high yield of BC at optimal conditions [39–41], and this is achieved using the traditional HS medium for BC production. BC amount is 6.26 g/L after 7 days obtained at static cultivation [41], in the 50/50 R-TS—HS medium. Therefore, the strain *Gluconaceto-bacter sucrofermentans* B-11267 in agitated culture conditions without any pre-treatment or addition nitrogen source and highly acidic by-products of the alcohol in order to, production of bacterial cellulose (BC).

5.1 Isolate of *G. sucrofermentans* B-11267 (bacterial strain) from kombucha tea

Bacterial strain was prepared in a test tube suspension using 1 ml of the suspension tea and then 9 ml of 0.9% sodium chloride (w/v) was added. The groups of different dilutions (10 to $1x10^{-6}$) were synthesized by solution from sterilized saline. About 0.1 ml of each dilute is taken on a media (agar (15 g/L), yeast extract (10 g/L), glucose (100 g/L), and calcium carbonate (20 g/L)). The plates were put in an incubator at 28°C for 3 days. After 3 days, we see a clear zone around with colonies, and they were selected and moved into glass vessel having 10 ml of Hestrin and Schramm (HS) medium [39].

The strain of produced positive cellulose is in the liquid medium, and the media for BC production include the following (g/L): yeast extract (5), citric acid (1.15), glucose (20), peptone (5), and disodium hydrogen phosphate (2.7), at pH 6, which is called "Hestrin and Schramm medium" (HS): thin stillage (TS) without pH adjustment, pH 5 and pH 6, cheese whey without pH adjustment, pH 4.96, pH 3.95; in autoclave for 20 min at 120°C [39].

5.2 Effect of thin stillage (TS) to produce bacterial cellulose

In this work, they investigated *G. sucrofermentans* B-11267 in agitated culture utilizing thin stillage (TS) to result in bacterial cellulose and whey without any modification as food sources to lower the manufacturing costs. For comparison, the HS medium was used. We find that increasing the bacterial cellulose using thin stillage after 3 days the yield is equal (6.19 g/l) and this amount of BC is approximately three times than that produced by HS (2.14 g/l). In the whey medium, the yield of bacterial cellulose is equal (5.45 g/l) after 3 days of cultivation. However, the maximum rate of the production of bacterial cellulose is the first day during the growth of the bacterium [39].

5.3 Effect of pH on the production of BC from thin stillage (TS)

The production of bacterial cellulose was affected by pH of the culture medium with thin stillage. So the bacterium *G. sucrofermentans* B-11267 on TS gives BC amount [39] (6.19 g/L) at pH 3.95.

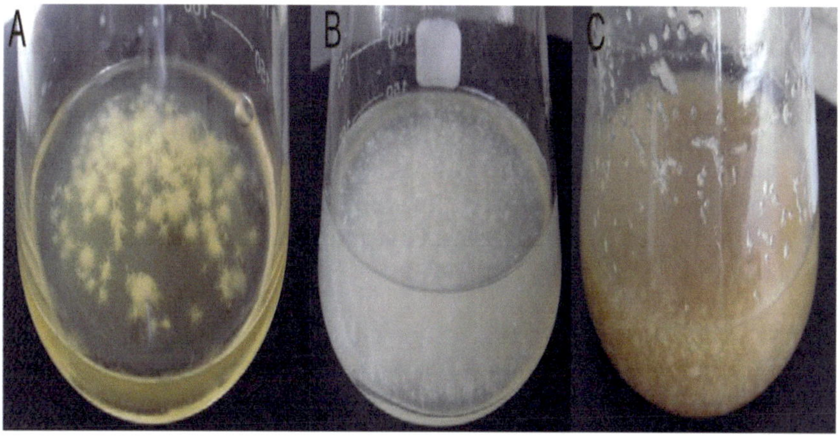

Figure 6.
Gluconacetobacter sucrofermentans B-11267 in agitated culture conditions using HS medium (A), whey (B), and thin stillage (C) to produce BC.

In **Figure 6**, it was shown that the collection of bacterial cellulose is formed in different sizes and shapes. It is clear that thin stillage (TS) and whey have the finer shape and homogeneous structures of BC produce in B, and C images compared to standard HS medium.

From this study, it was indicated that in the alcohol and dairy industries, such whey and thin wheat stillage are used as crude materials for producing bacterial cellulose (BC). So BC can be produced by using thin wheat stillage to give high yields and good quality. Further, the strain *G. sucrofermentans* B-11267 explained here yields a large amount of BC at acidic pH media [39].

6. Effect of lignosulfonate on the produce of bacterial cellulose

The acid-sulfite pulping of wood is the waste product of lignosulfonate. However, this lignosulfonate has more properties, such as high dispersive and adhesive abilities, and can be used as a soil stabilizer [42]. In this work, they study production and structure of bacterial cellulose. However, the presence of several types *of Gluconacetobacter*

Strain parameter	HS medium			HSL medium			Yield ratio (HSL/ HS)
	Yield (g/30 ml)	Gluconic (μg/20 μl)	Final pH	Yield (g/30 ml)	Gluconic (μg/20 μl)	Final pH	
10,245	.1344	5.43	2.98	.2178	2.59	4.03	1.62
13,693	.2376	7.41	3.4	.4896	2.61	4	2.06
13,772	.2628	5.73	3.5	.3432	2.11	4.06	1.31
13,773	.3036	6.84	3.55	.4866	2.7	4.25	1.6
14,815	.114	8.83	3	.234	2.94	4.06	2.0
15,237	.1272	8.94	3.01	.312	2.5	4.11	2.45

Table 3.
The weight of BC (mg) per 30 ml of culture medium after cultivation for 7 days.

xylinus (=*Acetobacter xylinum*) strain utilizes only 15237, IFO 13693, ATCC 10245, 13772, 14815, and 13773 in the presence of lignosulfonate. This study evidences this all types of strains produced BC with improving nearly 57% with added (1%, w/v) lignosulfonate, and the higher crystallinity index cellulose, which means amorphous region in the presence of lignosulfonate was relatively lower. These data indicate the high BC yield due to the damage of gluconic acid in the presence of lignosulfonate. That lignosulfonate contains (antioxidant) polyphenolic compounds. **Table 3** shows an increase in BC yield in the presence of 1% lignosulfonate using Hestrin-Schramm (HS) medium [42].

7. Biosynthetic pathway of BC in *Gluconacetobacter xylinus*

The BC biosynthesis in the presence of enzymes as catalysts is used by the following steps: (a) Glucose is converted to glucose-6-phosphate using glucokinase as enzyme; (b) the glucose-6-phosphate produced from glucose is isomerized to glucose-1-phosphate; (c) uridine diphosphate glucose (UDP-glucose) is produced from glucose-1- phosphate by UDP-glycose pyrophosphorylase; and (d) finally, glucose uridine diphosphate produced glucan chains [43, 44] (cellulose) using cellulose synthase enzyme.

After this, the chains of glucan were arranged in parallel and crystallized to construct the microfibrils region; then, the microfibrils are aggregated to form bundles of cellulose fibers [13, 45], by complete washing to remove culture medium, and for purification of the yield, colorless, odorless, and tasteless, they have obtained the BC in the form of a gel. So, its presence has several applications in our life for this gel product [45]. The following mechanism shows the bacterial cellulose synthesis pathway in *G. xylinus.*

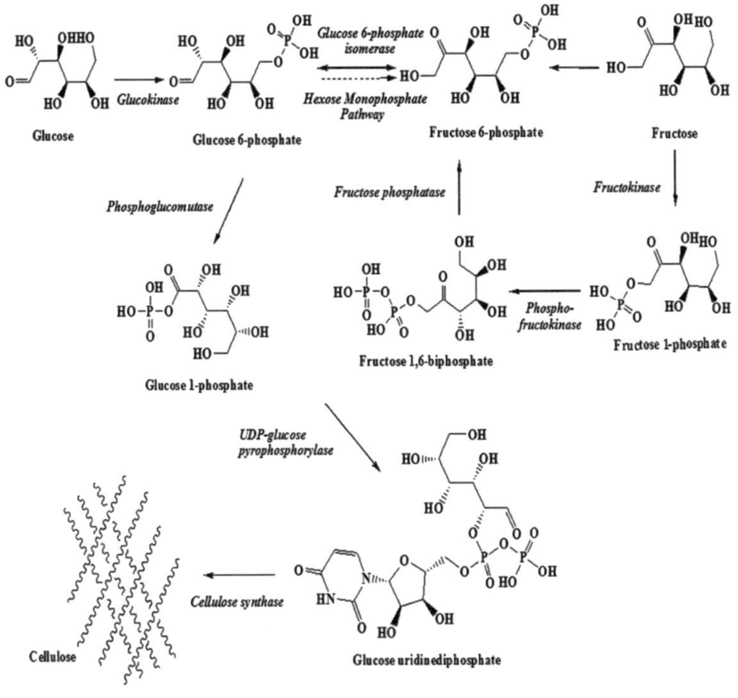

8. Applications of bacterial cellulose (BC)

Due to its biocompatibility, hydrophilicity, biodegradability, and nontoxicity properties for cellulose, it is the most abundant biodegradable material and has been widely used in medical applications, such as wound dressing, tissue engineering, controllable drug delivery system, and blood purification [46]. The bacterial cellulose has better properties compared to plant cellulose, such as higher crystallinity (80–90%) [47], water absorption capacity [48], and a higher degree of polymerization (8000) [49]. Finally, the characteristic properties of bacterial cellulose included wound care, biosensors, tissue engineering, drug delivery, and diagnostic [50], which are the medical applications of BC composites.

According to the above-mentioned properties, several applications of the BC, such as cosmetics, foods, and drug delivery, are present.

8.1 Bacterial cellulose in medical applications

BC-based materials are used in biomedical applications [51]. Due to the ideal structure, biocompatibility and sustainability of BC have led to many studies and prompted its application in a variety of fields, such as medicine [47, 52–54]. Nowadays, BC-based materials are mostly utilized in the medical field, including in wound healing materials, artificial skin and blood vessels, scaffolds for tissue engineering, and drug delivery [23, 55–58].

8.1.1 Bacterial cellulose for wound dressing

In medicine, field dressing material is used as a band aid or as a large bandage. By properties such as biocompatible, sterile, porous, and flexible, it is also used as a protective surface for firefighters, who are often exposed to burns (**Figure** 7) [59]. Such material allows the breathability of wounds, to prevent the formation of scabs and scars that must make different treatments. Also, this reduces pain, protects the skin from various infections, and does not cause loss of body fluid. Advantages of bacterial nanocellulose produced by Biofill® are rapid pain relief, close adhesion to the wound, noticeably rapid dressing, reduction in wound size after surgery, wound

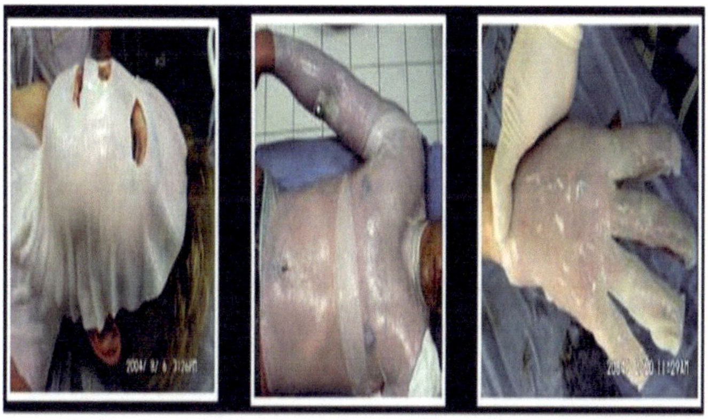

Figure 7.
Dressings made with bacterial cellulose that is imposed on burned tissue [59].

control (transparency), reduction in infection rate, cost reduction, and treatment time. The only drawback of such bacterial nanocellulose is that it cannot be used in the more mobile parts of the body since it is less elastic [60].

Whatever the reason, the high hydrophilic properties of bacterial cellulose and the fact that it never dries make these good properties because they indicate that better wounds heal faster and must be moisturized. Bacterial cellulose has a favorable property for use as skin tissue scaffolds and wound dressings [60].

On the market today, few BC-based commercial products are available for wound dressings of cavities, abrasions, and also as chronic venous ulcers and healing for burns. The structure of BC has porosity. So, it is possible to impregnate drugs in its structure, and this property can develop bacterial cellulose properties, for example, antimicrobial activity. But also, it is incorporates some elements, such as copper and silver [61].

8.1.2 Bacterial cellulose applications in tissue engineering and scaffold

Currently, the bacterial cellulose (BC) has full potentials to select as substrate material in tissue engineering. There are still many challenges to overcome the control porosities of BC scaffold by optimizing culture conditions, which is challenging, increasing BC degradation rate for specific applications and introducing functional groups to BC matrix [56, 62, 63]. So BC has physical modification, such as change porosities, crystallinities, and fiber densities and chemical modification in chemical structure to achieve this goal (TE, scaffold) and the chemical modification can occur by changing the carbon source of BC. It can modify BC done by either chemical or physical processes to evaluate the effectiveness of the composite scaffolds in construct bone regeneration.

In order to evaluate the biological properties of the new BC scaffold for bone regeneration, some materials must be added, such as hydroxyapatite nanoparticles (HA - Ca_5 $(PO4)_3(OH)$), to utilize an additive in the BC culture medium [56], the new bone tissue takes 4-weeks post-implantation [64], and the bone defects were completely fill.

Table 4 shows the added *in situ* treatment of BC for TE applications, and additional materials additives must be controlled to prevent the microbial fermentation limit. Also, using *in situ* modification method, till now, the structure of BC nanofibers still needs to be addressed such as overlapping between BC fibril growth and the externally introduced additives.

Formula G. A: Gluconacetobacter; Acetobacter.

There are two methods for *ex situ* modification of bacterial cellulose (BC), which are as follows:

(i) Chemical modification: the chemical structure of BC polymer is cellulose. So, BC could be modified by phosphorylated compounds and then treated by grafting or reactions with crosslinking to give BC modification [80]. In this method, the bacterial cellulose has a hydroxyl group that can make a strong hydrogen bond between the BC. The films and tubes from bacterial cellulose (BC) are prepared with calcium-deficient hydroxyapatite (BC/Cd HA) for application in bone tissue engineering [81].
(ii) Physical modification: the porous of BC can be filled with solution or particle of suspension of absorbed molecules additives; in this study, impregnated of AgNPs (BC/AgNPs) by immersing into BC membrane, the silver prepared a BC membrane nitrate as precursor ($AgNO_3$) solution after which the impregnated silver ions (Ag^+) were reduced to Ag^0 particles [82], and the inhibition zone of using BC/AgNPs was

DOI: http://dx.doi.org/10.5772/intechopen.107021

Application	Bacteria	Additive material	Modified properties	References
Scaffold for TE	*A. xylinum*	Collagen	Roughness, stiffness color, thickness, crystallinity, and	[65]
Bone regeneration	*A. xylinum*	Polystyrene and optical	porosity	[66]
Wound dressing/artificial	*G. hansenii*	fibers	Porosity, and swelling behavior, fiber network, crystallinity,	[67]
Skin	*G. hansenii*	Sand dollar skeleton	and mechanical property	[64, 68, 69]
Vascular grafts	*G. hansenii/Gluconacetobacter*	Hydroxyapatite	Density, porosity, mechanical property, and thickness	[70–72]
Biofiltration	*xylinus*	Paraffin	Fiber diameter and orientation, thermal property,	[73]
Urinary reconstruction	*G. hansenii*	Glucose or dextrin	mechanical property, porosity, crystallinity, and	[74]
	G. xylinus	Potato starch	biocompatibility	[75]
	A. xylinium	Cotton gauze	Porosity, biocompatibility, and mechanical property	[48]
	A. xylinium	Aloe vera	Rehydration rate, surface area, water-holding capacity,	[76]
	A. aceti	Deacetylated chitin	biocompatibility, and porosity	[77]
	A. xylinum	Nanocrystals	Crystallinity, thickness, density, porosity, mechanical	[78]
	G. medellinensis	Chitosan and heparin	property, biocompatibility, and rheological	[79]
	A. xylinum	Carboxymethyl cellulose	Wicking ability, water absorbency, drying time, and porosity	
	Paraffin		Porosity, mechanical property, surface area, water vapor	
			permeability, water absorption capacity, and crystallinity	
			Antimicrobial activity and mechanical property	
			Porosity, roughness, surface area, and biocompatibility	
			Thickness, porosity, water retention, charge, and	
			composition	
			Porosity, biocompatibility, and mechanical property	

Table 4.
Summary of bacterial cellulose for TE applications for in situ modifications.

tested against *Staphylococcus aureus* and *Escherichia coli*. It can see the inhibition zone around the sample.

As an example of a scaffold, BC is loaded by bone morphogenetic protein-2 (BMP-2) to study the possibility of utilizing BC as a scaffold for bone tissue engineering. The localized growth factor delivery system has more bone formation and higher calcium concentration than blank BC scaffolds from 2 and 4-weeks post-implantations [83].

8.1.3 Bacterial cellulose in blood vessel and cartilage

The blood vessel is also field using bacterial cellulose in medicine application. However, bacterial cellulose can be prepared as nanocellulose film or sheet compared with organic sheets, such as polypropylene or cellophane, and has high mechanical strength, such as tear resistance and shape retention. These properties are better for artificial materials than other materials. Consequently, it can be made the prototype of blood vessels have a tube of 5–25 cm long [84].

Where the cartilage is wider in adults and children, this cartilage is made from bioactive material, such as bacterial cellulose. The high wide in this topic is the nose, an ear, and intervertebral discs using reconstructive surgery.

8.1.4 Bacterial cellulose as drug release

As the structure of bacterial cellulose contains hydroxyl groups and other good properties such as purity, crystallinity, porosity, and water-holding capacity, some polymeric compounds and BC have been studied for controlled drug delivery. The synthesis of nanocomposite of BC to optimize the controlled drug delivery is an important strategy for pharmaceuticals in order to achieve the drug-delayed release effects of BC. In some studies, the matrix of BC and polyacrylic acid (PAA) (BC-PAA) has been synthesized by polymerization initiated through electron-beam irradiation using various doses of radiation [85, 86]. In this case, pH, swelling rate, gel fraction, and gelling time of prepared hydrogel must be controlled. The composite hydrogels from BC-PAA are utilized for drug delivery with different contents of bovine serum albumin (BSA) as a model compound [85].

8.2 Bacterial cellulose in textile applications

The textile industry must be concerned with quality and environmentally friendly products. Various research studies are required for the development of this industry, including preparation, finishing, and dyeing as important factors for sustainable development and economically feasible for the population. So, a hydrophobic cellulosic finishing is needed because it has a wide range of applications, not only in conventional applications but also in functional applications, such as clothing, waterproof textile, stain resistant (oils), antimicrobial, soil release, and self-cleaning. The ideal cellulose fabrics for water repellency are hydrophobic fiber surfaces because they resist water, with some porosity that allows moisture transport for user comfort [87].

8.2.1 Bacterial cellulose as water repellent finishing of textile

The main goal of this study was to use water repellent bacterial cellulose (BC) to create a material with potential application in design, specifically in the textile industry.

Some additives, including the incorporation of a softener, were used to improve BC flexibility. This finishing bath contains a certain ratio of nearly 60% of commercial hydrophobic product, finishing agent, and six samples of bacterial cellulose (BC) in different concentrations to soluble in 0.5 ml of softener and after treatment, the samples of fabrics are then dried and cured in oven at 120°C for 1 minute. The contact angle can be measured by using bacterial cellulose in finishing bath, which means the more hydrophobic of samples in the presence of bacterial cellulose (BC). This is due to the presence of coating layers on the surface of textile to decrease the surface energy [88].

8.2.2 Bacterial cellulose applied as comfortable textile

In this study, bacterial cellulose as breathable and water impermeable depends in the preparation of nanocomposites, using two commercial hydrophobic polymers to treat as water/oil repellency and comfortable fabrics in textile, and Persoftal MS (polydimethylsiloxane) and Baygard EFN (perfluorocarbon) are used on bacterial cellulose (BC) membranes, by an exhaustion technique [89]. However, by incorporating the commercial hydrophobic material with porous of bacterial cellulose as network, the contact angle measured in the presence of finishing by bacterial cellulose alone, in the presence of softener polydimethylsiloxane, (S) and in the presence of hydrophobic compound (perfluorinated) (H), where the contact angle in the presence of bacterial cellulose alone is equal to (63.8°).

By evaluating water vapor permeability (WVP) and static water absorption (SWA), this property is the most important for the comfort textile. When the concentration of S and H composites in BC is increased, the WVP decreases when compared to BC alone.

8.3 Bacterial cellulose in food application

The bacterial cellulose is utilized in food because it is a dietary fiber and is known as a "generally recognized as safe" (GRAS) food by the US Food and Drug Administration (FDA) [90]. BC is widely applied in the food industry [91]. The nata-de-coco, a juicy and chewy dessert from the Philippines, is made from the material of bacterial cellulose. The raw material produced from BC is grown from coconut water with many carbohydrates and amino acids. After manufacturing, it can be cut into cubes and put in sugar syrup [13]. The Monascus-produced meat analog—the BC complex has the best property, such as dietary fiber, and the lowering of the cholesterol from Monascus as well as the non-animal origin. This BC makes the product that is a suitable substitute for meat products for humans with dietary restrictions [43].

8.4 Bacterial cellulose in cosmetic applications

The human body must use cosmetics substances to enhance some of the organoleptic characters [92]. Alternating the appearance of the person's body utilizes cosmetic products that are cleansing or beautifying the body parts and enhancing the attraction, without affecting the normal body functions or structure [92]. Bacterial cellulose has biodegradability, low toxicity, and ability to hydrate the skin, so this BC treats dry skin. The cosmetics are applied in the case of oil-in-water emulsion without any addition of surfactant. The bacterial cellulose has high water-holding capacity and good gas permeability, so it is used as an appropriate carrier for the active ingredients for cosmetic materials. The cosmetics materials can be used as moisturizers and have whitening ingredients, for example, hyaluronic acid and salicylic acid, kojic acid or ursolic acid, anti-wrinkling agents (e.g., exfoliator and polypeptides), growth factors, and a combination thereof or enzymes [93].

The mechanical properties of bacterial cellulose result in the adhesion force of the mask; therefore, the structure of the formulation includes hydrophilic and hydrophobic chemicals. So, these properties provide a better interface between the skin and the mask. Images of the masks adhered to the face and hand skin indicate the good attractive of the mask to the skin [94].

9. Conclusion

Bacterial cellulose (BC) or microbial cellulose (MC) is a bioactive substance with high water absorption, crystalline structure, high tensile strength, and biodegradability. However, bacterial cellulose has a wide range of uses, including biomedical, textile, paper, food, medication release, and cosmetics. So FTIR, XRD, SEM, and TEM are used to characterize the microbial cellulose production from *Acetobacter xylinum* using diverse waste as carbon and nitrogen sources, such as pineapple peel juice, sugar cane juice, dry olive mill residue, waste beer yeast, and wheat thin stillage. The concentration of sugar in the carbon source, the temperature and time of the strain's incubator, and the pH of the media all influence the production of bacterial cellulose.

10. Future of perspectives

Production of the bacterial cellulose industrially with high yield and green synthesis uses an industrial fermenter design to grow *A. xylinum* culture.

Production of bacterial cellulose from the environment wastes uses less water and energy.

The applications of bacterial cellulose are widen to meet various trends of the modern world.

Author details

Ahmed Amr and Hassan Ibrahim*
National Research Centre, Textile Research and Technology Institute, Giza, Egypt

*Address all correspondence to: hmaibrahim@gmail.com

References

[1] Esa F et al. Overview of bacterial cellulose production and application. Agriculture and Agricultural Science Procedia. 2014;**2**:113-119

[2] Helbert W, Cavaille J, Dufresne A. Thermoplastic nanocomposites filled with wheat straw cellulose whiskers. Part I: Processing and mechanical behavior. Polymer Composites. 1996;**17**(4):604-611

[3] Chawla PR et al. Microbial cellulose: Fermentative production and applications. Food Technology and Biotechnology. 2009;**47**:2

[4] Vitta S, Thiruvengadam VJCS. Multifunctional bacterial cellulose and nanoparticle-embedded composites. Current Science. 2012;**102**(10):1398-1405

[5] Huang Y et al. Recent advances in bacterial cellulose. Cellulose. 2014;**21**(1):1-30

[6] Somerville CJARCDB. Cellulose synthesis in higher plants. Annual Review of Cell and Developmental Biology. 2006;**22**:53-78

[7] Ioelovich M, Leykin AJB. Structural investigations of various cotton fibers and cotton celluloses. BioResources. 2008;**3**(1):170-177

[8] Zadorecki P, Michell AJJPC. Future prospects for wood cellulose as reinforcement in organic polymer composites. Polymer Composites. 1989;**10**(2):69-77

[9] Franz G, Blaschek W. Cellulose. In: Methods in Plant Biochemistry. Elsevier; 1990. pp. 291-322

[10] Fengel D, Wegener G. Wood: Chemistry, Ultrastructure, Reactions. Walter de Gruyter; 2011

[11] Saka S. Chemical composition and distribution, chapter 2. In: "Wood and Cellulose Chemistry". By David NS Hon and Nobuo Shiraishi. New York-Basel: Marcel Dekker, Inc.; 2001

[12] Yamada Y et al. The phylogeny of acetic acid bacteria based on the partial sequences of 16S ribosomal RNA: The elevation of the subgenus Gluconoacetobacter to the generic level. Bioscience, Biotechnology, and Biochemistry. 1997;**61**(8):1244-1251

[13] Iguchi M, Yamanaka S, Budhiono A. Bacterial cellulose—A masterpiece of nature's arts. Journal of Materials Science. 2000;**35**(2):261-270

[14] Nakagaito AN, Iwamoto S, Yano H. Bacterial cellulose: The ultimate nano-scalar cellulose morphology for the production of high-strength composites. Applied Physics. 2005;**80**(1):93-97

[15] Pecoraro É et al. Bacterial cellulose from Glucanacetobacter xylinus: Preparation, properties and applications. In: Monomers, Polymers and Composites from Renewable Resources. Elsevier; 2007. pp. 369-383

[16] Steinbüchel A, Rhee SK. Polysaccharides and Polyamides in the Food Industry: Properties, Production, and Patents. Wiley-VCH Verlag GmbH & CO. KGaA; 2005

[17] Sakurada I, Nukushina Y, Ito T. Experimental determination of the elastic modulus of crystalline regions in oriented polymers. Journal of Polymer Science. 1962;**57**(165):651-660

[18] Castro C, Zuluaga R, Putaux J-L, Caro G, Mondragon I, Gañán P. Structural characterization of bacterial cellulose produced by Gluconacetobacter

swingsii sp. from Colombian agroindustrial wastes, Carbohydrate Polymers. 2011;**84**(1):96-102

[19] Brown AJ. XLIII.—On an acetic ferment which forms cellulose. Journal of the Chemical Society, Transactions. 1886;**49**:432-439

[20] Doi Y, Steinbüchel A. Biotechnology of Biopolymers: From Synthesis to Patents. Wiley-VCH; 2005

[21] Ramanaka K et al. Effect of various carbon and nitrogen sources on cellulose synthesis by Acetobacter xylinum in world. World Journal of Microbiology and Biotechnology. 2000;**16**(3):245-248

[22] Heo MS, Son HJJB. Development of an optimized, simple chemically defined medium for bacterial cellulose production by Acetobacter sp. A9 in shaking cultures. Biotechnology and Applied Biochemistry. 2002;**36**(1):41-45

[23] Klemm D et al. Bacterial synthesized cellulose—Artificial blood vessels for microsurgery. Progress in Polymer Science. 2001;**26**(9):1561-1603

[24] Hestrin S, Schramm MJBJ. Synthesis of cellulose by Acetobacter xylinum. 2. Preparation of freeze-dried cells capable of polymerizing glucose to cellulose. The Biochemical Journal. 1954;**58**(2):345

[25] Arahal DR et al. Value of recN Sequences for Species Identification and as a Phylogenetic Marker within the Family "Leuconostocaceae". 2008

[26] Dellaglio F et al. Description of Gluconacetobacter swingsii sp. nov. and Gluconacetobacter rhaeticus sp. nov., isolated from Italian apple fruit. International Journal of Systematic and Evolutionary Microbiology. 2005;**55**(6):2365-2370

[27] Brown RM, Willison JH, Richardson CL. Cellulose biosynthesis in Acetobacter xylinum: Visualization of the site of synthesis and direct measurement of the in vivo process. Proceedings of the National Academy of Sciences USA. 1976;**73**(12):4565-4569

[28] Yamanaka S, Sugiyama JJC. Structural modification of bacterial cellulose. Cellulose. 2000;7(3):213-225

[29] Gomes FP et al. Production of bacterial cellulose by Gluconacetobacter sacchari using dry olive mill residue. Biomass and Bioenergy. 2013;**55**:205-211

[30] Kurosumi A et al. Utilization of various fruit juices as carbon source for production of bacterial cellulose by Acetobacter xylinum NBRC 13693. Carbohydrate Polymers. 2009;**76**(2):333-335

[31] Hong F, Qiu KJCP. An alternative carbon source from konjac powder for enhancing production of bacterial cellulose in static cultures by a model strain Acetobacter aceti subsp. xylinus ATCC 23770. Carbohydrate Polymers. 2008;**72**(3):545-549

[32] El-Saied H et al. Production and characterization of economical bacterial cellulose. BioResources. 2008;**3**(4):1196-1217

[33] Keshk SM, Razek TM, Sameshima K. Bacterial cellulose production from beet molasses. African Journal of Biotechnology. 4 September 2006;5(17):1519-1523

[34] Kongruang S. Bacterial cellulose production by Acetobacter xylinum strains from agricultural waste products. In: Biotechnology for Fuels and Chemicals. Springer; 2007. pp. 763-774

[35] Lin D et al. Production of bacterial cellulose by Gluconacetobacter hansenii CGMCC 3917 using only waste beer yeast as nutrient source. Bioresource Technology. 2014;**151**:113-119

[36] Liu M et al. Optimization of extraction parameters for protein from beer waste brewing yeast treated by pulsed electric fields (PEF). African Journal of Microbiology Research. 2012;**6**(22):4739-4746

[37] Yang C-Z et al. Study and applications of technology aboutbreaking yeast cell wall. Letters in Applied Microbiology. 2006;**7**:268-274

[38] Ge H-J et al. Gluconacetobacter hansenii subsp. nov., a high-yield bacterial cellulose producing strain induced by high hydrostatic pressure. Applied Biochemistry and Biotechnology. 2011;**165**(7-8):1519-1531

[39] Revin V et al. Cost-effective production of bacterial cellulose using acidic food industry by-products. Brazilian Journal of Microbiology. 2018;**49**:151-159

[40] Wu J-M, Liu R-H. Thin stillage supplementation greatly enhances bacterial cellulose production by Gluconacetobacter xylinus. Carbohydrate Polymers. 2012;**90**(1):116-121

[41] Wu J-M, Liu R-H. Cost-effective production of bacterial cellulose in static cultures using distillery wastewater. Journal of Bioscience and Bioengineering. 2013;**115**(3):284-290

[42] Keshk S, Sameshima K. Influence of lignosulfonate on crystal structure and productivity of bacterial cellulose in a static culture. Enzyme and Microbial Technology. 2006;**40**(1):4-8

[43] Ullah H, Santos HA, Khan TJC. Applications of bacterial cellulose in food, cosmetics and drug delivery. Cellulose. 2016;**23**(4):2291-2314

[44] Ross P et al. Cellulose biosynthesis and function in bacteria. Microbiological Reviews. 1991;**55**(1):35-58

[45] Lin S-P et al. Biosynthesis, production and applications of bacterial cellulose. Cellulose. 2013;**20**(5):2191-2219

[46] Sindhu KA, Prasanth R, Thakur VK. Medical applications of cellulose and its derivatives: Present and future. Nanocellulose Polymer Nanocomposites. Wiley Blackwell; 2014;**20**:437-477

[47] MAS KS. Bacterial cellulose production and its industrial applications. Journal of Bioprocessing & Biotechniques. 2014;**4**(2):1

[48] Saibuatong O-A, Phisalaphong M. Novo aloe vera–bacterial cellulose composite film from biosynthesis. Carbohydrate Polymers. 2010;**79**(2): 455-460

[49] Dahman Y. Nanostructured biomaterials and biocomposites from bacterial cellulose nanofibers. Journal of Nanoscience and Nanotechnology. 2009;**9**(9):5105-5122

[50] Moniri M et al. Production and status of bacterial cellulose in biomedical engineering. Nanomaterials (Basel). 2017;**7**(9):257

[51] Gallegos AMA et al. Bacterial cellulose: A sustainable source to develop value-added products–a review. BioResources. 2016;**11**(2):5641-5655

[52] Mohite BV, Patil SVJB, Biochemistry A. A novel biomaterial: Bacterial cellulose and its new era applications. Biotechnology and Applied Biochemistry. 2014;**61**(2):101-110

[53] Evans BR et al. Palladium-bacterial cellulose membranes for fuel cells. Biosensors and Bioelectronics. 2003;**18**(7):917-923

[54] Cannon RE, Anderson SM. Biogenesis of bacterial cellulose.

Critical Reviews in Microbiology. 1991;**17**(6):435-447

[55] Czaja W et al. Microbial cellulose—The natural power to heal wounds. Biomaterials. 2006;**27**(2):145-151

[56] Czaja WK et al. The future prospects of microbial cellulose in biomedical applications. Biomacromolecules. 2007;**8**(1):1-12

[57] Wang J et al. Bacterial cellulose: a natural nanomaterial for biomedical applications. Journal of Mechanics in Medicine and Biology. 2011;**11**(02):285-306

[58] Czaja W et al. Biomedical applications of microbial cellulose in burn wound recovery. In: Cellulose: Molecular and Structural Biology. Springer; 2007. pp. 307-321

[59] Shah J, Brown RM. Towards electronic paper displays made from microbial cellulose. Applied Microbiology and Biotechnology. 2005;**66**(4):352-355

[60] Grzegorczyn S, Ślęzak A. Kinetics of concentration boundary layers buildup in the system consisted of microbial cellulose biomembrane and electrolyte solutions. Journal of Membrane Science. 2007;**304**(1-2):148-155

[61] Kucińska-Lipka J, Gubanska I, Janik HJPB. Bacterial cellulose in the field of wound healing and regenerative medicine of skin: Recent trends and future prospectives. Polymer Bulletin. 2015;**72**(9):2399-2419

[62] Stumpf TR et al. In situ and ex situ modifications of bacterial cellulose for applications in tissue engineering. Material Science Engineering C Material Biological Applications. 2018;**82**:372-383

[63] Semlali A et al. Cigarette smoke condensate increases C. albicans adhesion, growth, biofilm formation, and EAP1, HWP1 and SAP2 gene expression. BMC Microbiology. 2014;**14**(1):61

[64] Saska S et al. Bacterial cellulose-hydroxyapatite nanocomposites for bone regeneration. International Journal of Biomaterial. 2011:2011

[65] Luo H et al. Preparation and characterization of a novel COL/BC composite for potential tissue engineering scaffolds. Materials Chemistry and Physics. 2008;**110**(2-3):193-196

[66] Rambo C et al. Template assisted synthesis of porous nanofibrous cellulose membranes for tissue engineering. Materials Science and Engineering: C. 2008;**28**(4):549-554

[67] Barreiro A et al. Sand dollar skeleton as templates for bacterial cellulose coating and apatite precipitation. Journal of Materials Science. 2010;**45**(19):5252-5256

[68] Romanov D et al. Nanotextures of composites based on the interaction between hydroxyapatite and cellulose Gluconacetobacter xylinus. Glass Physics and Chemistry. 2014;**40**(3):367-374

[69] Grande CJ et al. Nanocomposites of bacterial cellulose/hydroxyapatite for biomedical applications. Acta Biomaterialia. 2009;**5**(5):1605-1615

[70] Feldmann E-M et al. Description of a novel approach to engineer cartilage with porous bacterial nanocellulose for reconstruction of a human auricle. Journal of Biomaterials Applications. 2013;**28**(4):626-640

[71] Zaborowska M et al. Microporous bacterial cellulose as a potential scaffold for bone regeneration. Acta Biomaterialia. 2010;**6**(7):2540-2547

[72] Brackmann C et al. In situ imaging of collagen synthesis by osteoprogenitor cells in microporous bacterial cellulose scaffolds. Tissue Engineering. Part C, Methods. 2011;**18**(3):227-234

[73] Stumpf TR et al. Enriched glucose and dextrin mannitol-based media modulates fibroblast behavior on bacterial cellulose membranes. Materials Science & Engineering. C, Materials for Biological Applications. 2013;**33**(8):4739-4745

[74] Yang J et al. In situ fabrication of a microporous bacterial cellulose/potato starch composite scaffold with enhanced cell compatibility. Cellulose. 2014;**21**(3):1823-1835

[75] Meftahi A et al. The effects of cotton gauze coating with microbial cellulose. Cellulose. 2010;**17**(1):199-204

[76] Butchosa N et al. Nanocomposites of bacterial cellulose nanofibers and chitin nanocrystals: Fabrication, characterization and bactericidal activity. Green Chemistry. 2013;**15**(12):3404-3413

[77] Wang J, Wan Y, Huang Y. Immobilisation of heparin on bacterial cellulose-chitosan nano-fibres surfaces via the cross-linking technique. IET Nanobiotechnology. 2012;**6**(2):52-57

[78] Orelma H et al. Affibody conjugation onto bacterial cellulose tubes and bioseparation of human serum albumin. RSC Advances. 2014;**4**(93):51440-51450

[79] Bodin A et al. Tissue-engineered conduit using urine-derived stem cells seeded bacterial cellulose polymer in urinary reconstruction and diversion. Biomaterials. 2010;**31**(34):8889-8901

[80] Rouabhia M et al. Production of biocompatible and antimicrobial bacterial cellulose polymers functionalized by RGDC grafting groups and gentamicin. ACS Applied Materials & Interfaces. 2014;**6**(3):1439-1446

[81] Zimmermann KA et al. Biomimetic design of a bacterial cellulose/hydroxyapatite nanocomposite for bone healing applications. Materials Science and Engineering: C. 2011;**31**(1):43-49

[82] Yang G et al. Antimicrobial activity of silver nanoparticle impregnated bacterial cellulose membrane: Effect of fermentation carbon sources of bacterial cellulose. Carbohydrate Polymers. 2012;**87**(1):839-845

[83] Shi Q et al. The osteogenesis of bacterial cellulose scaffold loaded with bone morphogenetic protein-2. Biomaterials. 2012;**33**(28):6644-6649

[84] Avery N et al. Collagen cross-linking in porcine M. longissimus lumborum: Absence of a relationship with variation in texture at pork weight. Meat Science. 1996;**42**(3):355-369

[85] Amin MCIM et al. Synthesis and characterization of thermo-and pH-responsive bacterial cellulose/acrylic acid hydrogels for drug delivery. Carbohydrate Polymers. 2012;**88**(2):465-473

[86] Halib N, Amin MCIM, Ahmad I. Unique stimuli responsive characteristics of electron beam synthesized bacterial cellulose/acrylic acid composite. Journal of Applied Polymer Science. 2010.**116**(5):2920-2929

[87] Dankovich TA, Hsieh Y-LJC. Surface modification of cellulose with plant triglycerides for hydrophobicity. Cellulose. 2007;**14**(5):469-480

[88] Araújo S et al. The role of technology towards a new bacterial-cellulose-based

material for fashion design. Journal of Industrial and Intelligent Information. 2015;**3**:2

[89] Fernandes M et al. Development of novel bacterial cellulose composites for the textile and shoe industry. Microbial Biotechnology. 2019;**12**(4):650-661

[90] Shi Z et al. Utilization of bacterial cellulose in food. Food Hydrocolloids. 2014;**35**:539-545

[91] Azeredo H et al. Bacterial cellulose as a raw material for food and food materials packaging applications. Frontiers in Sustainable Food Systems. 2019;**3**:7

[92] Hasan N et al. Application of bacterial cellulose (BC) in natural facial scrub. International Journal on Advanced Science Engineering and Information Technology. 2012;**2**(4):272-275

[93] Lin Y-C et al. Cosmetic Composition Containing Fragments of Bacterial Cellulose Film and Method for Manufacturing Thereof. Google Patents; 2015

[94] Pacheco G et al. Bacterial cellulose skin masks—Properties and sensory tests. Journal of Cosmetic Dermatology. 2018;**17**(5):840-847

Section 2

Recycled Textiles

Chapter 4

Progress of Recycled Polyester in Rheological Performance in Molding, and Economic Analysis of Recycled Fibers in Fashion and Textile Industry

Wei Tiancheng Wei, Yu Sun and Eunkyoung Shim

Abstract

In textiles, in particular wearable technology structured, battery-illuminated electronic fabrics are moving toward to both electrical and esthetic aspects of printed LED (PLED) textiles. It is on one dimension that have had questioned an economic resolution of extensional methods as for battery-charged interior materials, and also has gained a more general questions on how to develop its recycle both yarn and fiber as traditional raw goods in together with any recyclable electronic parts or graphene, carbon nanotube contained components with that textile materials. Furthermore, recyclable assembled electronic parts back to the renewable materials are continuous moving in the low-energy, high-reusable rate evaluation in the lifecycle assessment (LCA) of them. Specifically, during the de-sulfurization and decoloring in the post-production processing in the quick removal of active carbon fiber or nanoparticle coating on surface of fiber-porous geo-matrix could advance the renewing production efficiency. In assumption of low-energy and high-conductible transformation from off-market polyester with dyes or metallic yarns for polyester in e-textiles, recyclable conductive graphene/microfibers/composites are articulated, as far as industrial lifecycle management of braided, fibers, and ultra-high-density polyethylene has impeccable performance in the high mechanical property, medium rheological expansion over molding process, and high-yield strength as in the following sustainability in the wearable garment.

Keywords: polyester, lifecycle assessment, electrochemical textiles, 3D woven composites, production-consumption recycle

1. Introduction

In England, there are small-to-medium size quantities of fashion workshops that have launched their collaboration in recycled yarns. Such campaign involves fashion and textile supply chain from manufacturers, designers, distributor, and end use market consumers. It is estimated to recycle over 43% of second-hand clothes from Europe market [1] and over 269 million tons of urban-disposable plastic products were produced in an annual growth rate of 3.4% in the recycled polyester yarn market segments [2]. On the one side, there are concerns on eco-toxicity of consumption of waste textiles, fashion fabrics and packaging materials, on the other side, processing those products containing regenerated fibers such as viscous chitin, model, chitosan by using chemical spinning and wet spinning are available methods in reduce toxicity and reduce water consumption. As recycled thermoplastic package such as wiping waste materials, yarn fibers extract from felts, web, or even indoor carpet materials, which extend to grow at least in market capacity of 550, 000 to 900, 000 tons per year in UK, and over 1.7 billion to 3.5 billion global market segments as in global [3, 4]. Based on a new research of global recycled polyester filament yarn market, the food safety and pharmaceutical management, global soil contamination, and water pollution from textile synthetic yarns have been in account for nearly 10% of petroleum raw material as global recycled economy. This chapter is going to discuss about usage of polymer chain degradation, specifically fostering feedstock from bottle chopping to the antibacterial targeted to the global recycled polyester [5]. Polyester-recycled grids, films, laminated products in addition to prementioned polyester yarns are produced by wet spinning, and dry spinning is also included in this chapter as supportive various industrial semi-finished products in the supply chains.

There has discussion over relationship between increasing polymeric such as olefin-derivative, epoxy using glass-fibers and carbon-fiber, that production capacity must consider both the raw material prices and producing capacity of plants. Those virgin materials in production of petroleum derivative plastics in wearable electronics that are composed of recyclable virgin polymers such as polypropylene, could be prepared the lower cost in plants from both improving labor efficiency and inputing on-site recycling capacity for manufacturers [1, 6]. In terms of sustainability and hygiene quality improvement by using antioxidant and thermal rheological treatment in the relevant equipment, it has been tested at non-crystallized epoxy AEp-1, AEp-2, and AEp-3 showing viscosity in thermal rheological treatment in the relevant equipment, it has been tested at non-crystallized epoxy AEp-1, AEp-2 and AEp-3 showing viscosity of 7.3, 6.7 and 5.8 PA.S. This means that experimental results tend to have the same rheological behavior in as all three perceive same monomer polymerized epoxy antioxidant blend. It is found that cured AEP series thermosetting epoxy shows a room-temperature melting behavior for AEp-1 and AEp-2 as a liquidate status of non-crystalline polymers, while AEp-3 tends to while AEp-3 tends to show antioxidant component dependent crystallization temperature reducing over curing. It is found that AEP series thermosetting epoxy with standard temperature cure process tending to lower its crystallization temperature AEp-1 and AEp-2 while non-crystallized is due to larger enthalpy region for AEp-3. Liquidate transform of lower viscosity AEp-3 could result in the molding methods as the high. Using such acetal epoxy as the multi-use recyclable carbon composite materials could well maintain the mechanical and thermal properties of the materials. This research by Ningbo Chinese Science Institute by S. Ma, J. Zhu et al. have advocated new benzyl cyclic acetal epoxy with low viscosity [7]. This research proposes the carbon fiber recyclability in the three diol forms of acetal epoxy applied the carbon fiber-recycled process (CFRP) that well synthesizes the elongational tenacity, young's modulus, and strain at breaking factors. In another research paper by Q. Ma et al., it also supports he regenerated carbon fiber composite materials (regenerated CFRP) using the low viscosity monomer for uniform monomer morphological cross-link formed in the cured process [8]. In this chapter, a method of using cone and plate viscometry is for measuring the capillary *in situ* polymer melting and piston compressing modeling for the initial phase thermosetting material degrading. Their find is the one branch of recycle epoxy monomer with vacuum-assisted resin transfer molding (VARTM) process. Their finding proves there is a relationship between structure-property of epoxy monomers from experiment to the simulated computation about the acetal epoxy condensation formation.

A renewable bioresource especially lignin-derived vanillin is one other example of degradation of diacetyl epoxy resin reuse after distracting carbon fibers from the conventional bisphenol (BPA) epoxy resin. In a network, processing of cellulosic cotton materials is carried out by which internal jet dye tank provides a solution of dyeing tank to treat the florescent dyes in the previous process procedure. In the article by Sophie, at the viewpoint of sustainable fabric, they are proposing fabric by applying synthetic, fossil-fuel derived fabric as a direction for the garment-wrapped fabrics, which adds a condensation chain reaction in the elastomer production of algae derivative from wood pulp, bamboo, shell-extracted concentrates-producing fiber applied in the furniture, and decorated fabrics sustainable in the highlighted sustainable fabrics [9]. Also, the recycling cycle takes a short time so that right after the chip waste yarns are either chopped into small chunks or then feed through the vacuum moving belt, with a multi-ingredient separation at the seizures; those inconsistent materials are replaced with some of easily processible materials. As articles usually undergo three main steps in the post-recycling of polyester terephthalate-based recycling, in specific background procedures have adopted a continuous testing in the recycling system. Based on lifecycle assessment, there are a few blank spaces on the scenario of the intrinsic shifting efficiency from pre-shredding forms to the post-spun using wet forming or gel forming into the spinable materials. On the other hand, advanced multinational techniques have been applying the transboundary of semi-crystalline or amorphous low-density polyester in its subsequent production in the recycling products and its shifting with virgin products.

In the last decades, there is a growing concern of implementing the recycled polyester-chopped semi-finished goods, as return-to processing reduces the amount of chemical fluorescence content. Before life period of article comes till the end, there is always an immediate remembrance of that Japanese from primary family consumption to the school education emphasizing their automation in the recyclability of the products, which usually turns the virgin products in its primarily purified terephthalic acid form in a reaction that extends chain polymerization. This has cost extra Yen in the capacity of production because such excess production capacity could increase the post-processing cost in the corresponding LCA lifecycle assessment [10].

Recyclability of conductive fillers composited especially lithium polymer composite materials, carbon nanotube, graphene, and silicon carbide become intrinsic categories in the post-usage life circle of multiwall carbon nanotube, single-wall carbon nanotube, conductive silicon rubber, or spherical luminance filler particles. Because there are several parts in typical geogrids-composited road constructive geomaterials, in the industrial grades of molten materials that had been able to decompose to high viscous polystyrene components. The nickel metal hydride (NiMH)-applied electronic-conductive lead-acid batteries are performed as the collection of second life, which have a grid-connected energy storage even after restoring. The conductive materials, applied as vehicle-to-grid and large energy transaction in the electrical vehicle, may be used for grid compartment use and their interplanar intricated silicon carbide will have the partial recycle as form of different parts. But basically Li-ion battery has got the second-life batteries mostly after electrical cycle in the charging and discharging process reaching to a minimum withhold of less than 33%. But based on the second life of energy storage, nylon substrate composite is used with carbide or high-temperature ceramic materials. Recyclability of batteries separator fiber made of bimetallic carbide Co3Mo3C, have high energy conversion over sulfer/polysulfide/sulfer with similar electrocatalyst as LiS battery in the previous research [11]. In this chapter, a recalculation of the average battery price compares the pre-recycled form with the post-applied application [12, 13]. This revenue shortens the battery while reducing the reversible energy exchangeable rate. The index of storage rate to the migrating status was the electron to optical conversion rate to match the conversion of electricity in the repeated electricity charging. If assuming the manufacturing cost of photonic silicon as well as the cost of silicon PIC shown from the European CMOS may reach to 5 billion euro for enabling the prototyping of low volume manufacturing, then the electronic manufacturing lab required a higher energy conversion rate for the battery separator recycling applications, which also emphasizes that requiring such a high transparency and high index contrast in the 1200–3500 wavelength of both silicon and silicone dioxide materials in the photonic platforms.

A lifecycle inventory refers to the experimental awareness of how there is much inexpensive production in the discomposing the interlayer adhesive force of reduced graphene oxide composite materials (GO-CM) in spite of material form of graphite powder, graphene-coated materials, and the porous-vaporized carbon fibers in the storage sectors. Because both the forms of material array will endure a heat processing change over the subsequent thermal degradation reaction, and especially undergo the process of chain scissoring, co-mixing undergoing both physically and thermal forces impose onto the materials, and the suitable composite materials are usually harder than the basic thermoplastic materials in the process of its form. Their estimated energy scales up from laboratory scale from 140 to 300 kW cubic meter of graphene-sulfur composite with nitrogen-doping processibility capacity of 1500–2000 cycles of charging and recharging [14]. The low conductivity of sulfur materials as additional conductive additives such as carbon black, carbon nanotube, or silicon carbide, and boron nitride materials as in active form of additive will be bringing the challenge such as its toxic eliminating in its lifecycle management as it emits to the post-production process. For example, the sulfur elimination and photonic deposition from the nanosphere molecules are taken effective in its following cell manufacturing from electrode to the electro/electrolyte bi-modem sonication and layer-interfusing in its decomposition process. Currently, there is at least a decrease in the packaging energy decreasing as under European Union advanced sulfur battery program that makes the average time KW consumption in that lifecycle assessment (LCA) of Li-S battery.

2. Methodology

2.1 Melt flow index

In the high viscous molding articles when processing requires estimating thermal bulging in general processing, for instance in coat hanger spinneret for distribution, it is selected from an index of duration time that allows for pigmented polymer material preheating in the feeding system, which is followed by a continuous flowing from barrel to ambient air with a 10-minute counting for the material. Standards for characterizing the melt flow property of high flowability have melt flow index -MFI ≥ 50 g/10 minutes; for example, low-density polyethylene (LDPE) produced from the palletized materials in whatever assessment of middle degree of polymerization needs to be compared based on the flow amount of collected materials in that time window.

2.2 Viscometer

Time sequence interval recording of viscosity in poise as a unit has intrinsic viscometry measurement to consolidate the correlation between thickness uniformity with the shear acceleration. It is defined as shear rate that has derivative from viscosity to the unit acceleration interval. First, if assuming forming from period is important as the non-Newtonian usually involves with the shear rate increases as soon as the shear rate $\dot{\gamma}$ undergoes the nonlinear shear thinning behavior. In this perspective, this viscometer instrument is composed of three major situ-responsive numerical sensors to interpret the data from the capillary gadget, Rheometer-Rosand capillary rheometer RH7-2. In preparing procedures, two preheating sections are used to pressurize materials withholding materials inside the barrel. As preheat up to 230°C, the capillary is at the end of the barrel output with a constant mass of molten polyolefin at 24:1 of barrel, and a cone-entry front sits at the end of barrel. Any molecular crumble inside a 1-mm pinhole expels coiled olefins by the shear flow, which is called the viscoelastic flow, where the outward bulginess indicates pressure within the pinhole supreme high, especially for the polymer with a melt flow rate from 25 g/10mins to 36 g/10mins. The apparatus is first preheated up to targeted barrel temperature. The fiber samples of approximately 20-30 g were then loaded into the cylinder and tamped down by compressing in a twin-extruded process where extra bubbling squeezed extra bubble out. The sample was primarily preheated in the cylinder for 4 min to let the fiber samples melt; in the interval between the first and the second heating, a compression with 200Mpa pressure imposed a mechanical consolidation to the molten polymer. Then, pieces are cut by scissors without any molten-expanded room for it to squeeze fibers, and that molten polymer is attached to the barrel surface nozzle. It also withstands the compression force within the barrel but back-up by pressure from the piston to consolidate polymer materials.

2.3 Spectrophotometer

This spectrophotometer is applied in most of the applications that require the adjustment of pigment CIE light reflective index for the interior decoration plastic forms. In the specific types of spectrophotometer, measurement evaluation of CIE (a, b, L) includes x-y axis, which represents yellow-green and red-blue two-array indexes. In addition to that, L lightness mainly captures the third dimension of array to associate the brightness of comparison color plate in Daylight D65 light reflectance spectrum of molded polyester in the wavelength sensitive, specifically photon absorbance domain. Under batch of polyester woven or nonwoven structure, warp, and the structure array of yarn as interlayer of samples, as exampled one emission number varied based on output of data. Using this instrument, a template of cards is defined as intermediate to represent laboratory standard color plates with plain felt-like structure stands as a control board. Without surface lacquer print or vinyl backing light from spectrophotometer, light source will be transmitted across the randomly distributed polyester yarns without any excitation at the particle-treated yarn surface or the emission shied from the backing.

The coherent intercorrelation in the partial incident light imposes an intermediate film in filtering out-light. In this decorated vinyl-plasticized compounded materials in a series of post-molding processing, the light-modulated colorants may impose a color tint into remolded plastic materials. In such way, the intrinsic viscosity can be adjusted for a limited kneading within twin-screw extruder for dispersion of colorant inside. This intrinsic light refractive effect will be replenished by the interfusion over the dispersed over the colorants. In the relevant articulated models of PDMS chains, a contact silica surface is used for controlling the agglomeration existed in the paste control of those [15].

2.4 Fluorescence microscopy

Fluorescence microscopy integrates into forms of light, particularly with certain wavelength range that deals with the shift from high-frequency small wavelength to low-frequency larger wavelength. And such amount of energy emitted as fluorescence can further control the spectrum of emission energy in the spectrum, so that the collection of emitted fluorescence provides a screening of specimens with the light illumination property for the magnification and observation. In the splitting over the scatter screen over the condenser, this excitation light will be first illuminated when the low-frequency light entangled with the beam in its splitting the one-tenth of millimeter over the coherency filtering [16]. Using digital darkroom for testing a fluorescence microscopy that in-printing images of digital formats, the scattered particles are sputtered on the disks for the digital pixel detection in both crosswise and longitudinal manner. Over the number of techniques, LCD projectors that combine computer-aided software with the ink-jet printer provide such a photographic printing in which copies of high pixel printed on the samples were dye-sensitive to the exposure to the light source. In the microscopic level, tissue is used for objective lenses to clean for next few steps of fluorescent observation. Resolution is as follows:

$$R = 0.61(\lambda/NA) \quad (1)$$

where λ is the incident light wavelength, and N is the number of completed scatter pattern in the illumination considered as a filter with barrier for lamp exciting on the diagram, but not including the half-dissipated paths, while those half patterns are by light emitting on the blocks owning to Rayleigh scattering A.

3. Textiles perspectives in 2D and 3D assembling including spinning for reusability

To reveal fashion addressing the issue in both production and consumption, several technical indexes to review the technical in the wearer personal identification of rare purchaser, occasional purchaser, or frequent purchaser were discussed in the article by A. S. McGrath [17]. This textiles and fashion-related go-to-market end users are not only refereeing to individuals in the age from 18 to 25 who occasionally purchases for formal wear, leisure wear, or sportswear, they also have been within a group in age from 26 to 32 who emphasizes on their respondent to the human well-being of customer involvement in the satisfaction of quality-related problems in their purchasing [17, 18]. But in the domain of environmental responsibility and customer preference in the purchasing of valuable goods, the sustainable solution also guides the manufacturer of polluted materials, printed and dyed textiles, and mass manufacturing of clothing to take the resource as an index for efficiency aspects; in addition to that consumption in the aspects of technical improvements such as in the trendy, wearable products added with cultural and leisure aspects of fashion through the post-use and post-consumption link should reduce in considering the industrial disposable materials. In the reusable processing sectors, those defined as prototyping materials in the product design sectors, or the off-season products that encounter of accumulation in the warehouse for most factors related to price, speed of sales, or lack of enough wearable desire or lack of enough emotional needs, may be defined as un-needed or post-consumption goods. And both sustainable consumption and end-of-life-circle goods should be guided to either into developing country or to the second-hand market.

To overcome conventional economic theory related to the supply-need theory, choices should be based on mass media influences on the consumer un-biased interest of preference; meanwhile, it should take the effect of market-driven subjective environment into the psychological factor imposed onto consumer in being acted on consciousness of price-centered decision, materials-centered decision, or just human well-being as to the various of resource from external resource, specifically four factors that determine their consumption behavior, place, promotion, price, and person engaged. Compared to the cost of sustainable materials, energy-modulated spun-melt processing on the offset increases the cost of raw materials for manufacturers. The following questions will be solved. Starting at the point where there is an longevity of a goods with certain years of practical use value may take extra-long period for degradation in the environmental, unless it is sub-converted to sub-industrial goods [19]. Specifically, the end use of those manufactured raw materials can end up in the recyclable yarns and fabrics, or reproduced in the building sound-modulated wall sectors, or the manufactured goods applied in the packaging, suit-wrapping packaging, or cotton pads for either moisture regulation or absorption purpose. However in the LCA system, it is always lack of detailed explanation of HIGG Material Sustainability Index (MSI) especially in Sustainable Apparel Coalition (SAC) and handling of sustainable apparel coalition (SAC) Higg Materials Sustainability Index (MSI) methodology, or the environmental performance of wool textiles as well as the low product footprint category rules (PEFCR) in the general guideline and standards for the key assessment of the fabrics made from natural fibers such as cotton, organic cotton, cotton-polyester, linen yarns, canvas patches, silks [20].

3.1 Multiphase for photonic and electronic devices

Photonic silicone devices are in the optical medium for the micro-photonic intricate with the most used silicon photonic chips as built in the fabrication. Such fabrication is being installed in medical diagnosed, personalized medical devices, or the recovery medical fixture for aggregated tissue, artificial bones, or the silicone photonics implanted as the form of clinical readiness monolithic SiGe and heterogeneous embellished with micro-circular biosensors in the forms as previously mentioned methodology sessions as infrared resonance-silicone parts. Effective photonic component brings the similar components integrated in textiles. Demonstrated electronic devices in 2D electronic woven structure integrate the optical LED, microcontroller, printed circuit boards (PCBs), light-emitting diode, and array of LED. Such filaments include a flexible fiber/yarn abrasion for the wrapping of flexible modular circuits having designated as a sewn or the electronic-printed e-textiles for the purpose as LED-illuminated textiles. In maximizing reliability of interwoven disintegration parts for end-life circle recycling, the embedded LED light is either screen printed on the polymer film-coated woven yarns or using conductive adhesives as pasted on back of the woven belts. It is also delivered as an ultrasonic detective with a chip mounting as in the wire-bonded film onto the woven structure [21, 22]. Standing on such a screen-printed polymer conductors from Kapton with LED attached, in either woven structure where micro-LED units are incubated as a twisted unit, or in a printed copper-interconnected film, such unit compartments rely on the surface-raised interval arrays of thin copper lines that both integrated LED, PCB, conductive paste, and diodes as emitting two color arrays. Because the circuit printed on the woven structure could be relatively structural stable in any shearing or cross-sectional debonding, therefore, this type of application together with others as mentioned in Li-Ion battery (LIB) is incorporated on the cell manufacturing.

3.2 Conveying a screen-printed carding felt for doping and particle treatment

Carbonization of mesoporous silica deposition is done at the assembling as both micro-industrial particles and palletized specialty plastic, especially polyester, nylon 6, polyimide granulate over the water tank solidification. From the pre-screen of chopped pallets, those materials are categorized into high-melt flow index (MFI) for the product implication in the scenario of films, low-degradation temperature materials from polyethylene, polypropylene, and polylactic acid as produced in melt-blown processing. While the medium MFI was elaborated with monomer condensation between polyester, polyether, and polyester, polycarbonate and other thermoplastics contain nanoparticle forms of polystyrene granular, which configures the chemical oxygen demand removal by adding nanoparticles in extracellular polymeric substances for the wastewater treatment mechanism [23]. Extruder as equipment is applied in the analysis of shear viscosity, which may allow for melt viscous-incubated industrial chemical processing that is usually characterized as porosity or potential bulb in the curation of post-processing (**Figure 1**).

3.3 Viscous as index for mineral conductive fillers for spinning

As general exertion of product life evaluation, the mechanical performance of post-remolding process evaluates the flexural resistance and impacts resistance, while for the hydrostatic performance, recycled polyester yarns applied in the felt consolidation process that may be applied in filtration applications count the hydro-pressure, dust holding capacity, and static pressure in product development phase factors. Additional question on the viscosity modulation for the adaptive hydrolysis and organic substance compatible in the following biotreatment using the plastic molding, biaxial stretching, or blow molding process arises. The efficiency of filler particle packing is a predominant factor in increasing accurate loading volume in fibers. In this aspect, a diameter less than 1 micron up to 14 microns allows this empirical model to orient fibers in the web formation process of meltdown. And the web formation takes the high velocity of the air stream that directly impacts the particle such as titanium dioxide, calcium carbonate, and talc through die-channel distribution (**Figures 2** and **3**).

When going to the precipitated diffusive over the laser light, fine particle composition is one primary reason for forming pre-extrusion polymer and master batch kneading problems. Some of the most recent research have also pointed that a dimension gap of 10% large size in diameter should be equal to the exponential of finest 5% of size, so that the dendritic cluster will be first accumulated on top, so more good fillers could fill out those larger spaces left by stacking larger fillers. Indirect impact of changing concentration of particles will result in spinning temperature changes

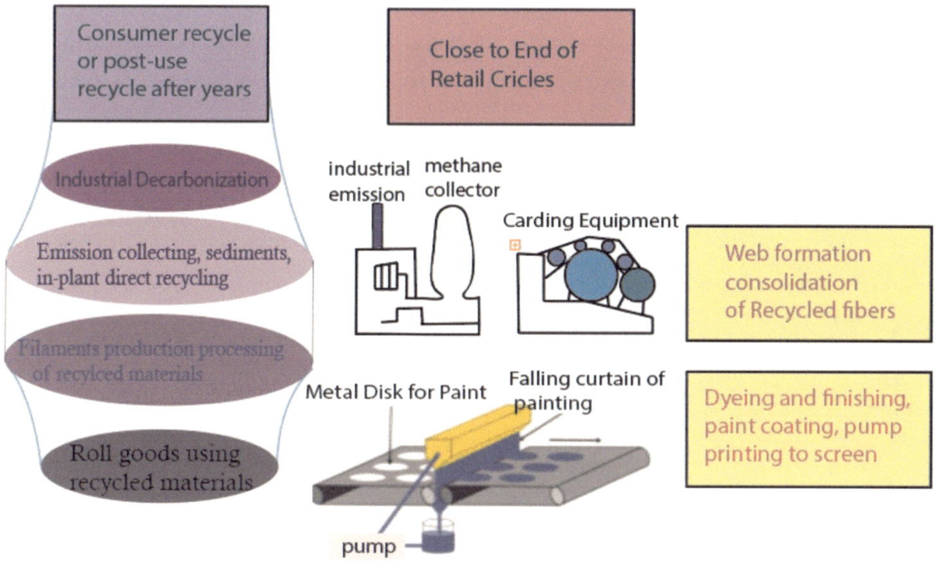

Figure 1.
Three phases of post-industrial recollecting, conveying in major processing including carding, laminating, and curtain painting to be reused for end of retail goods [24].

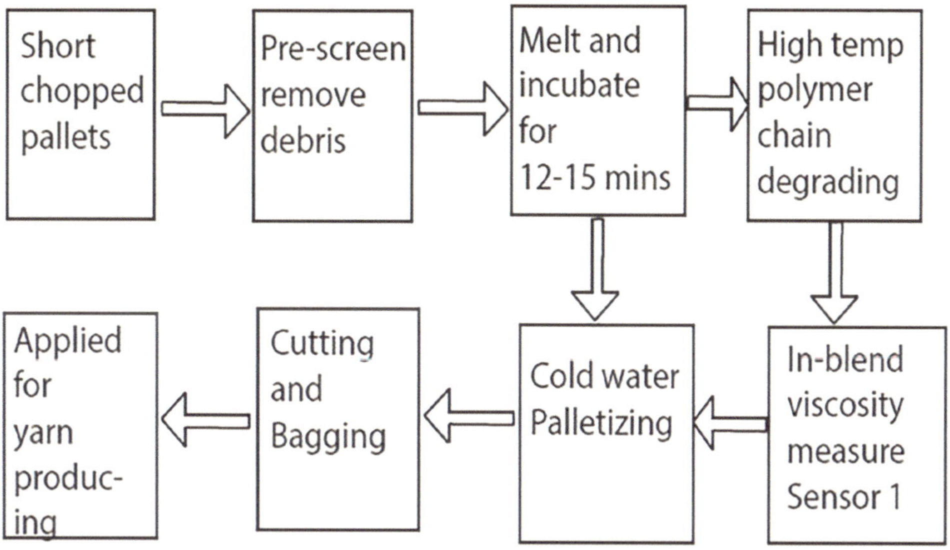

Figure 2.
The processing flowchart of screening and decomposing recycled plastics in the post-processing of industrial grade PET.

conducted from a different thermal conductivity resulting from inorganic fillers [26]. Usually, mono-disperse particles fillers have the lowest pack fraction at maximum. As to PCC-SF, as its particle size distribution is narrow and more petite than 0.6 μm, polymer evokes some agglomeration issue dependent on Φ_f. Theoretic results show that the ϕ_{max} is dependent on polydispersity, and minimum-to-maximum packing

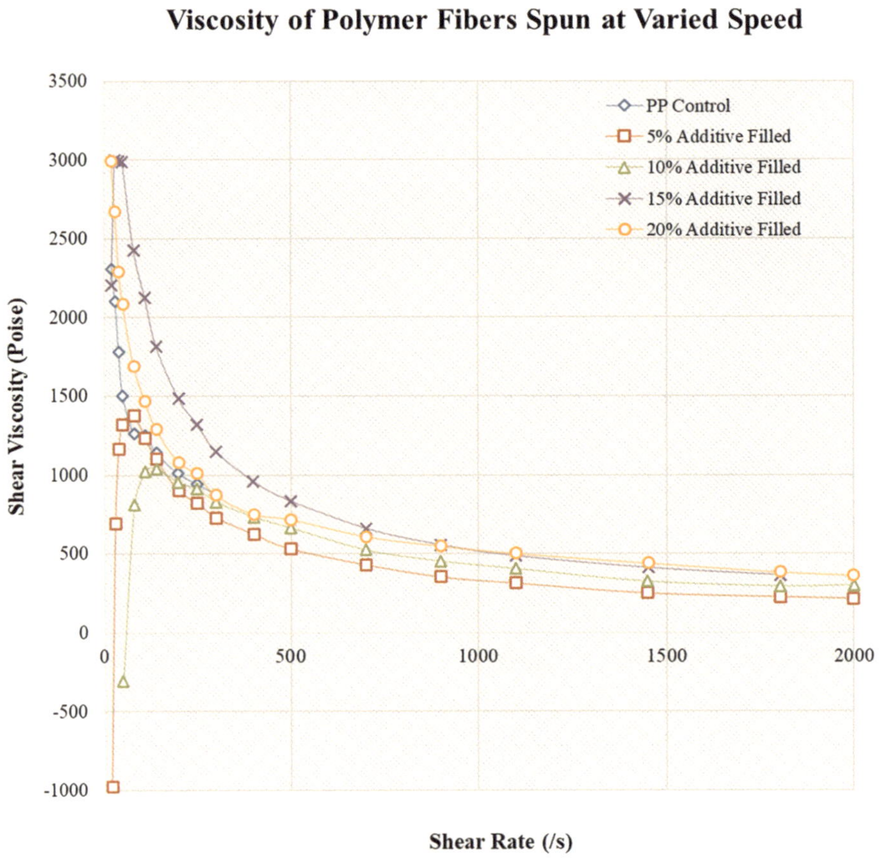

Figure 3.
Shear viscosity analysis of inorganic particulate filled polypropylene collected from spun bond aspirator drawn fibers as exponential shear viscosity to shear rate in 10 ~ 2000 s^{-1} range [25].

fraction is in a sequence as mono-dispersed < bimodal dispersed< tri-modal dispersed system, according to Genovese [27].

$$ln\,\frac{\eta}{\eta_0} = \frac{[\eta]\phi_n}{\sigma - 1}\left[\left(\frac{\phi_n - \phi}{\phi_n}\right)^{1-\sigma} - 1\right] \tag{2}$$

where σ: particle interaction coefficient and η: the suspension viscosity. η_0: an medium thickness, in numerical analysis. $[\eta]$: the intrinsic viscosity, which is more beneficial. ϕ_n: the particle packing fraction.

4. Reusable conductive composites containing graphene/PDMS/ microfibers

Obvious result of PDMS intact at a substrate coated with fluorescence or lithium in vapor-porous-carbon fibers may first form coherent force of chemical reaction, followed by a subsequent chain polymerization in the process. It basically allows for at least 12–15-minute incubation during which an alkaline solvent imposes any acid residue in the acid electrolyte left on the rechargeable materials, later it let the chain degrading with the reactor continued heating to a certain level. In the molten of thermoplastic materials, a transformation may bring the three in mechanism of material chain degradation, chain scissoring reaction, and replacement of substantial immobilized spinal chain molecules with the short chain forms. On the other side, there is the high throughput in both polyester processing and polyolefin extruding and the production of spun yarns in the flexible applications. Such applications have recently progressed in conductive fabrics containing carbon nanotubes, disposable electronic devices for blood pressure monitor, 3D printing fast-processing materials for fabricating post-recyclable biodegradable polylactic, and poly(lactic-co-glycolic) acid [28]. Such printed electronics are considered compatible with other thin layer of polyester/co-polyester hybrid bicomponent needle-punched materials as interior decorated fabrics containing sensor. And the robust of such stretchable elastic fiber contained structures that could build channel soft circuits or sensors used as inorganic/polymer binary phases in the display or printed wearable devices. The majority application area asks for stretchable performance that relies on certain composition between elastane fibers and inorganic filler composition, but overall integration methods through yarn upgrading and woven structure building especially spacer yarn constructed woven fabrics would gain more field-effect transistor integration with electronic materials for stretchable and mechanical durability.

Fabrics composed of the electronic elements as listed above are considered as knit, woven, braided materials, and it is a potential production-wise question on how the life circle of the commercial-size emission and waste management would have impacted on environment especially soil quality of landfill of those electronic battery, supercapacitor. In Matteo C. et al. series of research, they have applied three evaluations on the toxic of recycling of electronic that has energy storage functions into the environment: First dimension is using human body toxic on those devices containing capacitor, second dimension is on the greenhouse gas (GHG) and emission is generated by the chain scissoring during the decomposition of industrial graded materials including PTFE, HDPE especially graphene produced in the layered sheets, or the adhesive layer of conductive coating on top of the thermoplastic materials [12, 13]. The end life of battery recycled by discomposing of separator of battery from the lithium interlayer within the conductive medium is decarbonizing the greater percentage of carbon-based devices in the degradation process. Here in the consideration of third dimension, there must be value added by post-energy consumption of the electrical devices that have been added onto the materials. Generally, the value for consumption of a large amount of production cost and risk of associated with health-related processing in the plant would be considered as the additional environmental hazard that should be compared with basic plastic materials in thermoplastic sectors to the carbon nanotube and graphene composited plastics in the energy storage sectors [29].

4.1 Implying spacious filtering materials with adaptive structure

Biocompatible, specifically those garments that are related to the wound dressing, post-surgery recovery body extrudates treatment patches, and the bone conjecture is modulated by the sensor to detect the extracorporeal adaptation to the adjustment of partial harm recovery in the kinesiology application. And such on-body equipment for the exercising is wearable, which allows physician to monitor recovery from post-injury joint juncture. In the observation of spacious or liquid-absorbent materials used as dressing, growth-healing bandage, or the textiles, health problems are treatment functions for human body, such as the patches applied as pharmaceutical implementation over wound area, recyclable materials, especially textile-grafted base with either electrical sensors for monitoring or medicine-compatible hollow fibers that allow for adding of optical detective units around the impact area. In the response to multi-layer copolymer of polylactic acid and polyester terephthalate, copolymer co-extrusion into thin films laminates at least one layer of spacious superabsorbent or salt concentrated collective for the biological or wearable devices connected to human skins, body joints, or the multi-fibers made of nylon, of high-density polyethylene.

One discussion is applying the polyester stretching knitting fabrics, tubular-woven three-dimensional fabrics, or the high-density polyester/conductive fibers in warp yarns for multi-functions in electronics on wearable polyester, and bioimpedance and biocompatible polyester textiles containing electrodes, resistance, modules for athletes in their high-performance cold weather. The previous dissertation by Gunnarsson summarized a capacitive sensing using experimental models in the reported papers, as polyester yarns woven into conductive fabrics as of plain weave, twill weave, spacer weave on the primary applications [30]. Their hypothesis starting from the pressure sensory on the textile spacer materials could be applied as major changing of display on textiles according to their electric behavior. Bending for active wearable textiles for those environmental sensible application in medical, cable channels, or display electronics structured in spacer of woven especially interlacing polyester yarn, gaining a major barrier if yarn itself, is not susceptible to electronical charging, but the contacting resistance of monofilament yarns could bear certain degree of constriction resistance and pollution resistance [9, 31].

$$C = \frac{Q}{U} = \frac{A\varepsilon_0\varepsilon_r}{y} \tag{3}$$

where C is the capacitance of capacitor, Q is the amount of quantity of electric charge, U is the voltage, A, ε_0, ε_r, and y are the area of the plates, the permittivity of free space, the relative permittivity of the dielectric layer, and the distance of that two-plane parallel to each other. Thereafter, the three parts in the invention designated as spacer structure within have the interconnected woven as a separating field for either upper board A or bottom board. Meanwhile, the spacer yarn is bounceable that will be able to withstand the high capacitance for conductive purpose (**Figure 4**).

5. Conclusion: End use for integrating textiles for market-leading focus

In the another previous research based on a hypotheses of that textile business irrespective to the employment status for stores, factory stores, vintage shops, thrift stores, e-commerce sales, online resales tend to show a statistic correlation between consumers who either prefer to select the conscious of recycled material applied in the products or may or have high preference to the products in the apparel especially wears for workplace, performance, and those factors inside of region market. Because resources applied in the study have shown product evaluations for Chinese, Canadian consumers have different preference in the evaluative roles in different aspects of fit, style, prices, quality/workmanship, comfort, material/fabric, color, durability,

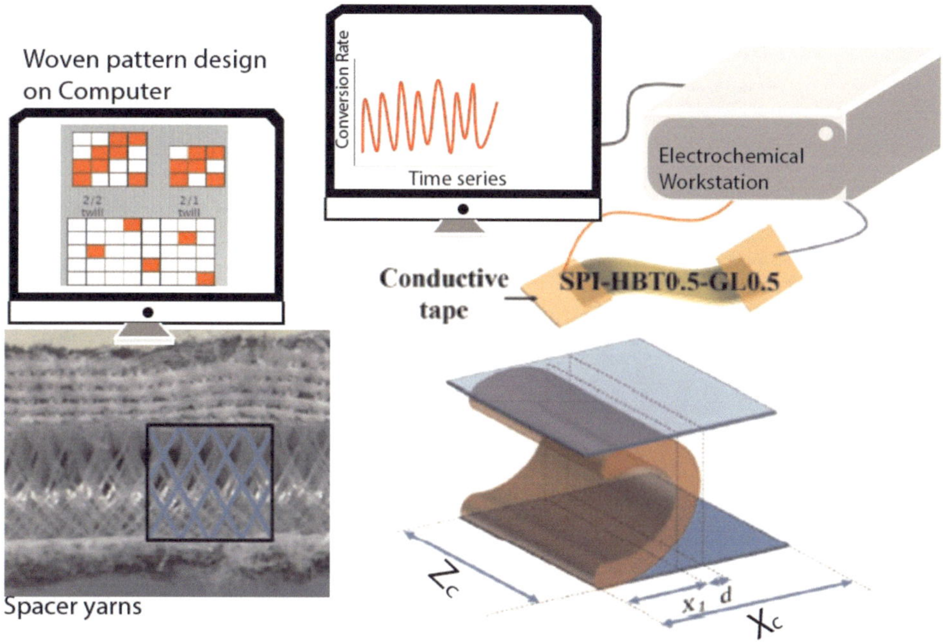

Figure 4.
The spacer structure and the lumped element model of a capacitor of three partial capacitances in the two parallel conductive boards, in which woven materials have a geometrical model, calculated based on the area of the plates, and the permittivity of free space in between the plates [31].

wardrobe coordination, ease of care, and brand name [9]. Also, the circular economy of potential future development of textiles is in either yarn or tread mill, dyeing plants, woven plants, post-treatment plants, no matter using their production capacity in the recycling of their industry waste. It also means enterprise including processing focused, or process to trade focused plants, and even the end of user market regulators to bring together multivariate variables in the modulation of textile and fashion industry, just for instance as in the prementioned research by (**Figure 5**) Watson et al. and Paul et al. [4, 20].

Our focus would be questioning on if next-generation textiles still stay on production—consumption-recyclable to industry circular economy system, the overall progress in the Open Design (OD) that should put secondary usage over excessive products in front of the purchasing directed garment wearing. Large consumption in recycled polyester materials in textile for both wearing and home decoration market has already built a common view on increasing the sustainable levels from traditional back-into rampant consumerism into the conservative consumerism added with the technology rebuilding over the value system. Circle for that constitutes both a detail-focused micro-manufacturing of small collection of accessories for fashion industry or an across-continent manufacturing of shirts that uses a labor-educated program in southeast Asia which projects material recovery, experimental scalable, customized program that has a proactive, system-based approaches to lower the garment impact by using a reduced processing emission from raw materials to the subsequent processing pollution management. Because the implementation of technology alone is not the eventual outcome and direction, it is better to use social education sectors for consumers to change their reliance on materials made of a cost-sensitive raw yarns, or

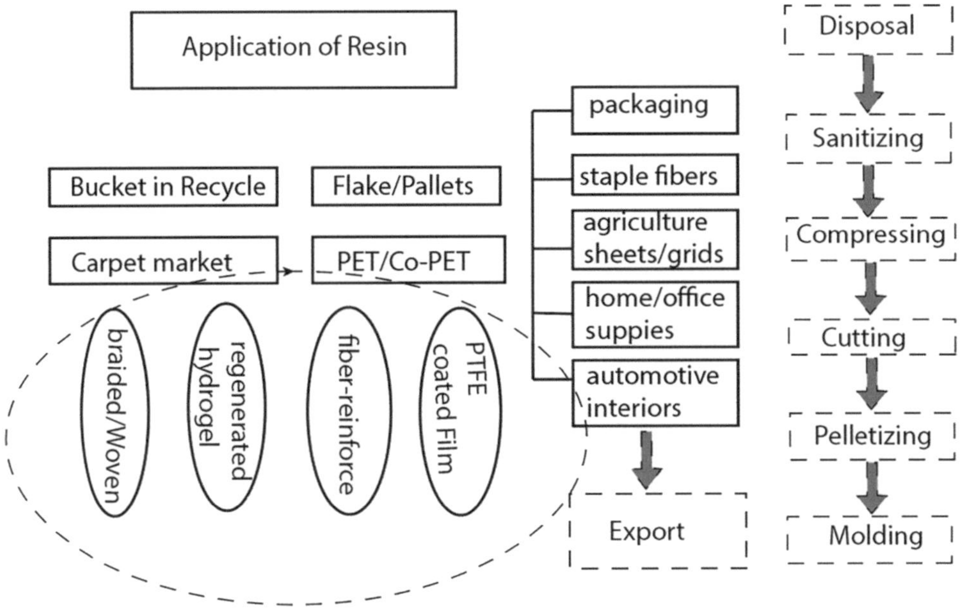

Figure 5.
Application and export of polymeric resins in its packaging forms and the industrial upgrade of form integration from packaging to the pelletizing forms where there are multiple export product application forms in the subsequent procedures.

sustainability-directed consumption. They also define the circular economy as the input of the shorter but more responsive grip in from designer, textile mills to the end-user [4, 18]. Post-processing batches of reusable yarns, treads, woven clothes, cellulose materials, post-treatment accessories, or parts of the needs are directed turning to the aggregation dispersed to solve fundamental needs in front but followed by the creation and multipurpose wearables for device-connecting purpose.

Author details

Wei Tiancheng Wei[1]*, Yu Sun[2] and Eunkyoung Shim[3]

1 Texafire Canada, Kingfa Science and Technology, USA

2 Mechanical, Automotive and Materials Engineering, University of Windsor, USA

3 Textile Engineering, Chemistry and Science, North Carolina State University, USA

*Address all correspondence to: wwei7@ncsu.edu

References

[1] Woolridge AC, Ward GD, Phillips PS, Collins M, Gandy S. Life cycle assessment for reuse/recycling of donated waste textiles compared to use of virgin material: An UK energy saving perspective. Resources, Conservation and Recycling. 2006;**46**(1):94-103

[2] Wang Q, Tang H, Ma Q, Mu R, Yuan X, Hong J, et al. Life cycle assessment and the willingness to pay of waste polyester recycling. Journal of Cleaner Production. 2019;**234**:275-284

[3] Karim MRB, Tahir D, Haq EU, Hussain A, Malik MS. Natural fibres as promising environmental-friendly reinforcements for polymer composites. Polymers and Polymer Composites. 2021;**29**(4):277-300

[4] Smith P, Baille J, McHattie L-S. Sustainable design futures: An open design vision for circular economy in fashion and textiles. The Design For Next Journal. 2017;**20**(1):1938-1947

[5] TMRGL. Recycled Polyester Textiles Market - Global Industry Analysis, Size, Share, Growth, Trends, and Forcast, 2019–2027. Transparency Market Research; 2018

[6] Amanda F, Maddy S, Tom C. E-textiles in clinical rehabilitation: A scoping review. Electronics. 2015;**4**: 173-203

[7] Ma S, We J, Jia Z, Yu T, Yuan W, Li Q, et al. Readily recyclable, high performance thermosetting materials based on a lignin-derived spiro diacetal trigger. Journals of Materials Chemistry A. 2019;7(3):1233-1243

[8] Li P, Ma S, Wang B, Xu X, Feng H, Yu Z, et al. Degradable benzyl cyclic acetal epoxy monomers with low viscosity: Synthesis, structure-property relationships, application in recyclable carbon fibers composite. Composites Science and Technology. 2022;**219**: 109243

[9] Rahman O, Fund B, Chen Z, Gao X. A cross-national study of apparel consumer preference and the role of product-evaluative cues. Journal of Marketing and Logistics. 2017;**29**:796-812

[10] Jun N, Minoru F, Yuichi M, Masahiko H. Life-cycle assessment of domesticand transboundary recycling of post-consumer PET bottles. International Journal of Life Cycle Assessment. 2010;**15**:590-597

[11] Zhang Z, Wang JN, Shao AH, Xiong DG, Liu JW, et al. Recyclable cobalt-molybdenum bimetallic carbide modified separator boosts the polysulfide adsorption-catalysis of lithium sulfur battery. Science China Materials. 2020;**63**:2443-2455

[12] Martinez-Laserna E, Gandiaga I, Sarasketa-Zabala E, Badeda J, Stroe D-I, Swierczynski M, et al. Battery second life: Hype, hope or reality? A critical review of the state of the art. Renewable and Sustainable Energy Reviews. 2018; **93**:701-718

[13] Munuera J, Britnell L, Santoro C, Cuellar-Franca R, Casiraghi C. A review on sustainable production of graphene and related life cycle assessment. 2D Materials. 2022;**9**:12002-12048

[14] Deng Y, Li J, Li T, Gao X, Yuan C. Life cycle assessment of lithium sulfure battery for electric vehicles. Journal of Power Sources. 2017;**343**:284-295

[15] Kim H, Miura Y, Macosko CW. Graphene/polyurethane nanocomposites

for improved gas barrier and electrical conductivity. Chemistry of Materials. 2010;**22**:3441-3450

[16] Jeff WL, Jose-Angel C. Fluorescence microscopy. Review. 2005;**2**:910-919

[17] McGrath AS. Fashioning sustainability: How the clothes we wear can support environmental and human well-being. Sustainable Fashion. 2012;**1**:1-24

[18] Lerpiniere C. Value definition in sustainable (textiles) production and consumption. In: Palgrave Studies in Business, Arts and Humanities. Cham: Palgrave; 2020. pp. 1938-1947

[19] Goworek H, Oxborrow L, Claxton S, McLaren A, Cooper T, Hill H. Managing sustainability in the fashion business: Challenges in product development for clothing longevity in the UK. Journal of Business Research. 2020;**117**: 629-641

[20] Watson K, Wiedemann S. Review of methodological choices in LCA-based textile and apparel rating tools: Key issues and recommendations relating to assessment of fabrics made from natural fibre types. Sustainability. 2009;**11**:3846-3862

[21] Abiodun K, Russsel T, Yang W, Helga NM, Menglong L, Dorothy H, et al. Integrating flexible filament circuits for e-textiles applications. Advanced Materials Technologies. 2019;**4**(7):1900176

[22] Wei Y, Jiang S, Li X, Li J, Dong Y, Shi SQ, et al. Green flexible electronics: Biodegradable and mechanically strong soy protein-based nanocomposite film for human motion monitoring. ACS Applied Materials & Interfaces. 2021;**13**:37617-37627

[23] Wei W, Hao Q, Chen Z, Bao T, Ni B-j. Polystyrene nanoplastic reshape the anaerobic granular sludge for recovering methane from wastewater. Water Research. 2020;**182**:116041-116054

[24] Borojeni IA, Gajewski G, Riahi RA. Application of electrospun nonwovens Fibers in air filters. Fibers. 2022;**10**:15-42

[25] Wei W, Shim E, Song Y, Pourdeyhimi B. Synthesizing heat transfer factor on thermal bonding structure of mineral added polypropylene spun-bond. Journal of Applied Polymer Science. 2022;**139**: 52138-52146

[26] Wei W, Shim E, Barnes W, Pourdeyhimi B. Structure-property relationship of melt spinning polypropylene fibers containing inorganic particulate $CaCO_3$ fillers. Textile Research Journal. 2021;**91**(11–12):1419-1435

[27] Genovese DB. Shear rheology of hard-sphere, dispersed, and aggregated suspensions, and filler-matrix composites. Advances in Colloid and Interface Science. 2012;**171**:1-16

[28] Bao Z, Xiaodong C. Flexible and stretchable devices. Materials Views. 2016;**28**:4177-4179

[29] Adetayo A, Runsewe D. Synthesis and fabrication of graphene and graphene oxide: A Reivew. Journal of Composite Materials. 2019;**9**:207-229

[30] Bilisik K. Multiaxis 3D woven preform and properties multiaxis 3D woven and 3D orthogonal woven carbon/epoxy composites. Journal of Reinforced Plastics and Composites. 2010;**29**(8):1173-1186

[31] Gunnarsson E. Conductive Fabrics for Textile Electronic Interconnections and Capacitive Sensing. Gothenburg, Sweden: Chalmers University of Technology; 2017

Chapter 5

Limitations of Textile Recycling: The Reason behind the Development of Alternative Sustainable Fibers

Gizem Celep, Gamze D. Tetik and Fulya Yilmaz

Abstract

The sharply increasing world population reveals the insufficiency of natural resources in meeting the needs of humanity, while creating a tendency to search for new resources. Textile products constitute one of the most basic needs of humanity and the consumption of textile products is also increasing due to the changing fashion sense, increasing population, and technology developments. Discovery of alternative or renewable energy sources, recycling of all kinds of materials, enhancing engineering methods and technologies used to make recycling effective, and trends like sustainable fashion that promote sustainability and take parts among the hot topics of this field. Recycling studies are also common in textile science. It is feasible to reduce the usage of natural fibers by utilization of recycled fibers. However, there are some limitations to textile recycling. These limitations led the development of new sustainable fibers and processes as alternatives to natural. In this context, most of the recycling and sustainability-based studies carried out in this field emphasized the indispensability of the subject, while neglecting a few points about limitations. Consequently, the limits of recycling in textiles and new fibers developed to overcome these limits are addressed in this chapter.

Keywords: textile recycling, textile wastes, limitations in textile recycling, textile eco-labels, sustainability certifications, sustainability

1. Introduction

Recycling is the process of converting materials from all kinds of waste to produce new products. Textile recycling implies the reuse and reprocessing of clothing scraps or any fibrous textile material [1]. All types of consumer or industry discarded textile goods are used as textile wastes for recovery. It is obvious that recycling, which has evolved into sustainability over time and its importance has been understood even in ancient times. It can be applied in many fields of the textiles as textile-to-textile (closed-loop) recycling or textile-to-nontextile (open-loop) recycling [2].

The demand for textiles and clothing is increasing day by day as a result of the increasing population, rising living standards, and the fast changing fashion trends [3–5]. Consequently, the amount of textile waste increases, and there are accumulations in landfills [4]. In addition to the consumption of high amounts of textile products, the packaging of these products also causes an increase in waste piles [6].

Textile and clothing waste causes environmental problems and deterioration of ecological balance. Therefore, the reclaim and disposal of waste clothing and textiles are important issues. Unfittingly and uncontrolled disposal of waste cause major problems [5]. The importance of recycling is considered in three subjects by Ref. 7 as economic, social, and environmental subjects [7]. Recycling of textile waste and diversifying the content of recycled raw materials could be a way to support the country's economy. The employment opportunities in the textile sector as in other sectors increase with well-run waste management. The recycling sector is an important supplier to many industries, and wastes are considered as cheap raw materials [8]. A wide variety of garment brand companies offer their products that contain recycled materials at certain rates, as a social responsibility issue in the market and to increase their prestige. It also adds profit to the company by paying less for recycled materials obtained from waste products.

Although there are several methods for the disposal of clothing waste, the most effective methods are recycling and reuse. Evaluation of waste clothing is very complex since clothes are made from different raw materials and may contain various accessories. Clothing may have many components such as labels, sewing threads, buttons, zippers, and interlining, and these components make the separation process difficult. Clothing recycling and textile recycling are two independent topics that are needed to be considered separately [5]. Textile wastes arise out of many production processes, such as fiber and filament manufacture, spinning, weaving, knitting, nonwoven, and clothing manufacturing [9]. In this context, textile wastes can be classified as pre-consumer and post-consumer wastes [10, 11]. Pre-consumer textile waste includes manufacturing waste from the processing of fibers, yarn, fabric, and nonwovens and clothing manufacturing [12]. Pre-consumer textile waste is generally seen as "clean waste" as it is released during the textile production process [13, 14].

When all these wastes are well managed, positive results emerge both in terms of providing economic gains via the recycling of materials and reducing the ecological damage to the world. Despite all advantages, there are recycling limits for all kinds of textile wastes. Not only for textile wastes but also for other solid wastes recyclability variates. Some types of wastes can be easily and well recycled, whereas some types cannot or can formidably be recycled [1]. These limits pave the way for the emergence of new recyclable fibers for the textile industry.

The purpose of this chapter is to present a systematic study for recycling of textiles mentioning the limits and alternative sustainable fibers. The content started with the history of recycling, continued with processes, standards, and certificates about textile recycling. Subsections of recyclable common textile fibers and new recyclable textile fibers are given in detail.

2. History of recycling

Recycling dates back to ancient times [15]. It is claimed that waste management and waste disposal processes date to BC in several references [16, 17]. Recycling is known as an efficient and effective solid waste management system [18]. In 4000 BC, silkworm wastes were used as protein source food in fish raising in China [19]. Scientifically, the foundations of recycling were laid in the 1980s [20].

When we consider textile recycling, it is known that it is as old as recycling in other fields. There are even references stating that it is one of the oldest fields, so textile recycling is called original recycling [21]. China hosted applications where recycled fibers from used clothing were obtained by hand carding and mixed with virgin fibers BC [14]. The textile recycling industry took its first steps in the thirteenth century [22]. In pre-modern societies, there were sustainability models based on the reuse and recycling of textiles [14]. For example, recycling has been done for years in India, both at the household and industrial level [23]. In the early and mid-1800s, reclaimed spin waste and rags were used for the manufacture of new products, and the invention of carbonization made it a unique technique to separate textile waste comprising of cellulose-based and wool fibers blend [22].

Environmental awareness concept had been newly introduced in the 1960s. The conscious interest of consumers and producers had just begun to turn to recycle at that time. Today, it is argued that this interest has evolved into sustainability [24].

Early academic studies conducted in the 1990s focused on presenting a model for the textile waste lifecycle [25]; detailing biological, physical, and chemical treatments of textile wastes [26]; determination of the number of sewn product manufacturers that support recycling in an American state [27]; the recyclability of post-consumer fibers, and market applications, while revealing the advantages of recycling [28]. After this decade, a positive acceleration was observed in the studies on both recycling and textile recycling. When "textile" and "recycling" terms are searched in a topic currently 1843 documents in WOS were encountered at all times. Moreover, 188,487 documents were encountered with the only term "recycling" at all times. The variation of the number of publications by years are given in **Figure 1** and in the first quarter of 2022, 41 documents were published about the textile recycling topic. As can be seen from the graph, the number of research on textile recycling has increased in parallel with the number of research on recycling over the years.

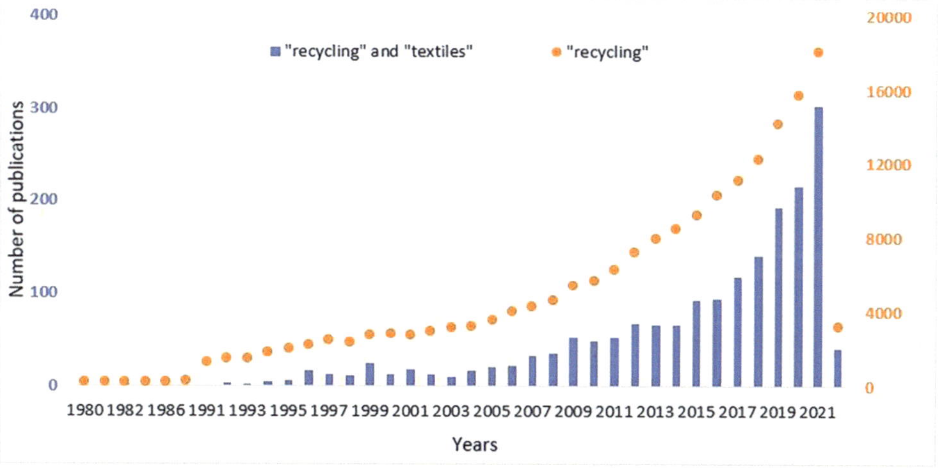

Figure 1.
The number of publications encountered with search terms "recycling" and "recycling and textiles" in WOS.

3. Processes in textile recycling

3.1 Recyclability of textile materials

Recycled fibers are used to make a variety of products. By producing yarns from recycled fibers, knitted or woven fabrics can be produced, or non-woven surfaces are obtained directly from these fibers. Recycled pre-consumer textile wastes are utilized in the construction, automotive, furniture, paper, and clothing industries. However, fibers obtained from pre-consumer textiles are used especially in coarse yarn production [29]. Many researchers studied about using pre-consumer waste and its conversion into a valuable product in the literature. Jamshaid et al. [30] span open-end rotor yarns from fibers in different blends reclaimed from yarn and fabric wastes. They evaluated the impact of various textile wastes on fiber and yarn quality. They underlined that the length and uniformity values of the fibers recycled from yarn wastes are better than those of the fibers recycled from fabric wastes. However, in terms of yarn manufacturing costs, it has been stated that yarns produced from recycled fabric/rag wastes are more economical than yarns produced from yarn wastes. The impact of cotton waste and various spinning conditions on rotor yarn quality was investigated by Halimi et al. [31]. The results showed that the quality of the rotor yarn is not affected by the addition of 25% waste in the first passage of the draw frame. Yilmaz et al. [32] produced yarns by blending the fiber wastes taken from the blow room, the carding and sucked in the draw frame, roving, and conventional ring spinning machines with the virgin cotton fibers at 5 different amounts varying from 5% to 40%. They emphasized that by designing machinery and process steps based on waste fiber type, it can be possible to produce yarns that are in comparable quality values and low cost.

The post-consumer textile wastes consist of clothing and home textiles that consumers no longer need for various reasons, such as damage, pulling on, or going out of fashion [14]. Contrary to pre-consumer wastes, post-consumer wastes are known as dirty and household waste [33]. Post-consumer wastes are evaluated with reuse and recycling techniques or incineration. The options to be applied to post-consumer waste vary according to many criteria such as the wear condition of the clothing, fiber content, and the technology of the recycling facilities [34]. The progress in recycling technology supports the sustainable disposal of waste clothing, and recycling is far more environmentally friendly and socially beneficial than incineration. In addition to this, technological advancements are required to produce upcycle products from waste clothing. Improvements in the collection and disposal of post-consumer textiles can be made with the application of environmental protection policies [5].

Post-consumer waste of sufficient quality is utilized as second-hand clothing by other consumers or sold to third-world countries. The volume of consumer waste is quite high, and clothes that cannot be worn again are shredded into fibers and used in new products, similar to pre-consumer wastes [6, 14]. The process of producing new clothing from post-consumer waste includes collecting waste, obtaining fiber from waste, and producing yarn by using a certain amount of blend in the yarn production stage [34].

When compared to original fibers, recycled fibers have different properties. The processes that the fibers are exposed to during the recycling process damage them and shorten their length. Fiber length is important factor in converting recycled fibers into yarn or producing nonwovens from these fibers, and the fibers must be long enough. Due to the short length of recycled fibers and the presence of non-fiber remnants such as fabric and yarn fragments, defining some quality parameters of

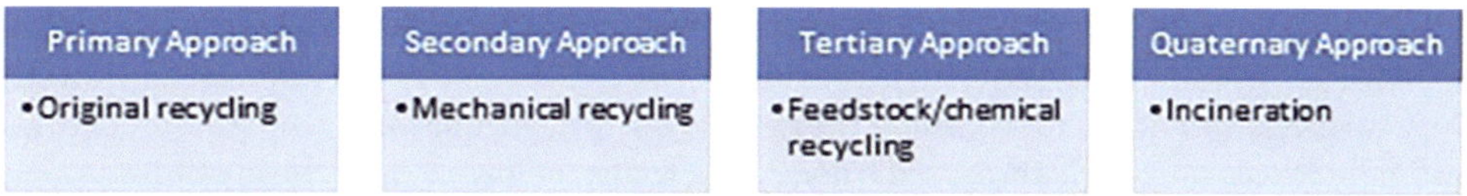

Figure 2.
Various recycling approaches.

these fibers is difficult. Fiber length, material break down degree, and fiber length distribution are three of the most widely analyzed properties of recycled fibers [35].

There are four different approaches to recycling (**Figure 2**) [36, 37]:

- Primary approach,
- Secondary approach,
- Tertiary approach, and
- Quaternary approach

Primary recycling is the most beneficial method, and in this approach, the product is recycled to its original form. This approach is also known as "original recycling." It is aimed at synthetic fibers such as PET (polyethylene terephthalate) and PA (polyamide) [37]. In this method, which can also be blended with the similar original raw material in order to increase the product quality, cleaned and pure scraps from waste are collected and recycled. In addition to the important advantages of this process such as being cheap and easy, it also has the disadvantage that the type of recyclable material is limited [38].

Secondary recycling is the process of converting waste into a product with different physical or chemical properties than the original [39]. Secondary recycling, which converts post-consumer wastes into raw materials, includes the collection and recycling process [40]. The content of textile waste, the degree of purity of the end product, availability, cost, and processing techniques are important factors for secondary recycling.

In tertiary recycling, known as feedstock recycling, wastes are separated into chemicals through pyrolysis, gasification, hydrolysis, and condensation [41]. Tertiary recycling, which is preferred for converting plastic wastes into chemicals, monomers, or fuels, utilizes clean and well-sorted pre-and post-consumer wastes [40].

Quaternary recycling is the use of heat produced by the incineration of fibrous solid wastes [42]. In summary, primary and secondary recycling usually involve the mechanical processes of industrial by-products and waste, while tertiary and quaternary recycling includes the pyrolysis and incineration of textile wastes for energy generation [41].

3.2 Recycling processes in textile

3.2.1 Physical processes in textile recycling

3.2.1.1 Mechanical process in textile recycling

Mechanical recycling is a low-cost and easy method [11], which is the preferred method for recycling a diverse variety of textile waste [43]. The recycling of post-consumer textile waste is generally carried out by mechanical recycling [42]. In the mechanical recycling technique, the fabric is broken down into fibers by cutting, shredding, carding, and other mechanical processes [44, 45]. Mechanical recycling machines gradually break the fabric into small pieces and make it fibrous, and these obtained fibers are reused in the production of yarn or nonwovens. In the mechanical recycling process, initially, wastes are sorted. Foreign components, such as metals and labels, are eliminated. After the fabric is cut into small pieces with rotary blades, it is separated into fibers by tearing [40].

Since garments are usually made from different raw materials, it is better to use pre-consumer waste instead of post-consumer waste in mechanical recycling. Fibers obtained by mechanical recycling from pre-consumer wastes such as denim scraps can be used to make higher-quality yarns. The length of the fibers recycled by the mechanical recycling process is short, despite the use of clean pre-consumer wastes [40]. The fiber length is shortened by the shredding/tearing process. The main reason for this is the friction between the fibers. Friction causes wear of fibers and melting of synthetic fibers. Lubricants are used to reduce friction between fibers during shredding and thus longer fiber lengths can be obtained [46]. In addition to the lubrication process, product quality is increased by blending original fibers with recycled fibers [45].

Recycled fiber properties such as length, fineness, and strength indicate the field the fibers can be evaluated in [47]. Good quality recycled fibers can be spun into fabrics, while lower quality fibers are used as decoration materials, construction materials [48], automotive components, insulation materials, and nonwovens [45, 47].

3.2.1.2 Thermal process in textile recycling

In thermal recycling, synthetic fibers are melted to be reshaped. The thermal recycling method is preferred for recycling synthetic fibers [48, 49]. Chips and pellets obtained by mechanical process from synthetic wastes are turned into fibers by melt extrusion [50].

3.2.2 Chemical recycling in textile

Chemical recycling, which is another method used in the recycling of textile waste, is the depolymerization of polymers or the process of dissolving polymers [2]. Polymers are converted or broken down into their original monomeric building blocks by chemical and biological methods [51].

Monomer and polymer recycling are the two forms of chemical recycling. The polymer chain is frequently degraded during polymer recycling. As a result, the quality of the recycled fiber decreases. In monomer recycling, original quality fibers are obtained. While monomer recycling is only used for synthetic fibers, chemical recycling is applicable to many textile fibers [49]. In addition to the chemical recycling of synthetic fibers, such as polyesters, polyamides, and polyolefins, in cotton and polyester blend products, the fibers can be chemically separated and then converted into new fibers [47].

3.2.3 Downcycling

Downcycling occurs when the quality and economic value [40] of a product obtained from recycling processes is lower than that of the original product [2, 50]. The use of recycled clothing and home textile wastes in agriculture and gardening

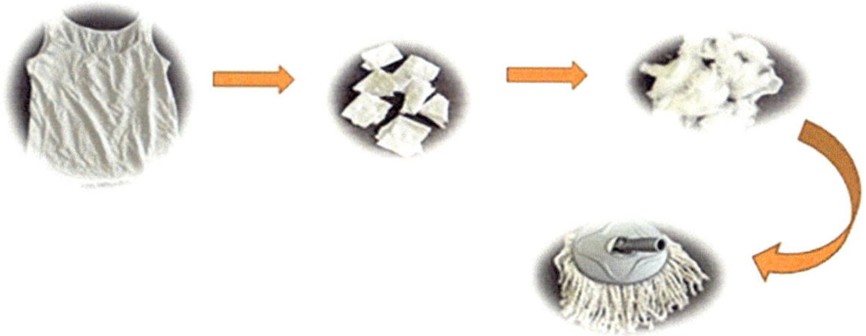

Figure 3.
Downcycling applications in textile.

products, decoration materials [48], insulation materials, low-quality blankets, and upholstery fabrics are the examples of downcycling (**Figure 3**) [2, 50].

3.2.4 Upcycling

When the quality of the recycled material is the same or higher than the original product, this process is called upcycling [2, 50]. Upcycling is a process in which existing resources are used and converted into more useful products. This environmental-friendly process is an important step for a zero-waste policy [52]. Within the scope of sustainability and circular economy, the production of raw materials such as cotton fibers and yarns from textile wastes with polymer and monomer recycling is an example of upcycling [2, 50].

3.2.5 Open-loop recycling in textile

Open-loop recycling is defined as the use of a product's raw material in a different production area. Secondary products obtained through open-loop recycling are generally destroyed after their lifetime [40]. The use of fibers obtained by recycling PET bottles in the textile industry (**Figure 4**) [40] and the usage of recycled textile fibers as insulation material in the construction industry are examples of open-loop recycling.

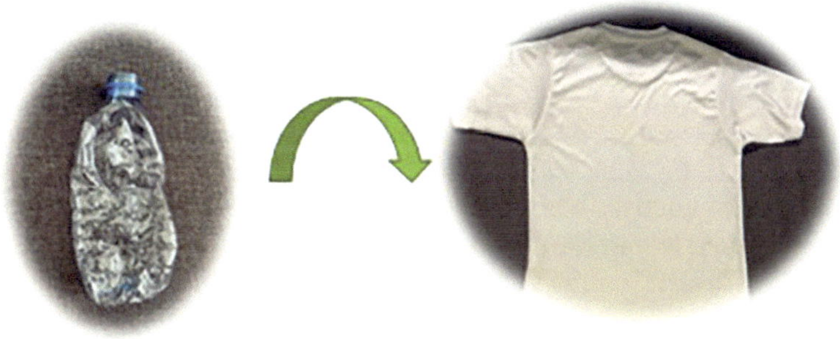

Figure 4.
Open-loop recycling.

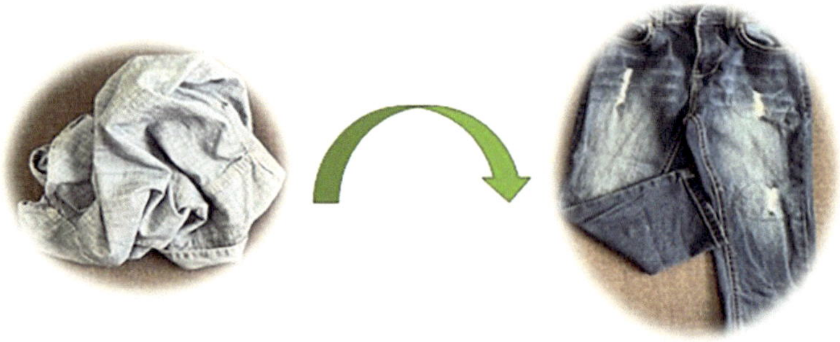

Figure 5.
Closed-loop recycling.

3.2.6 Closed-loop recycling in textile

The reuse of recycled textile waste in the textile industry is called closed-loop recycling [2]. The use of mechanically recycled pre-consumer or post-consumer textile waste in garment production is an example of a closed-loop recycling (**Figure 5**) [40].

4. Sustainable textile Fibers

4.1 Recyclable common textile fibers and limitations

The subject of recycling in textiles comes up with a lot of research based on the advantages created by the recycling process and with a limited number of studies based on recycling limits. In this subsection of the chapter, the limits of recycling in materials traditionally used in textiles are addressed.

Despite the approaches expressing that recycling is a process that only delays the conversion of waste to nature [53], several articles emphasized the importance of process development studies about the determination of recyclability limits [54]. Since it is impossible to apply a uniform recycling method for recycling all kinds of waste materials, different recycling techniques and their combinations have emerged over time [53]. For example, chemical recycling is raised in order to eliminate the limits in mechanical recycling [55]. As recycling can be classified as mechanical, chemical, thermal, and thermomechanical methods; each of them has numerous disadvantages in terms of the imperfections created on the recycled material. Considering these limits, alternatives purposed for the disposure of textile waste as anaerobic digestion, fermentation, composting, and acquisition of construction material [13].

An assessment can be made on the basis of fiber source for recycling limits. Based on the disadvantages, such as shortening or shredding the fibers created the fibers by each recycling cycle, it is stated in the literature that an average of 8 recycling cycles for synthetic fibers and an average of 5 cycles for natural fibers can be actualized [56, 57]. While the recyclability limits are more evident in natural fibers, the same rule is not valid for the fibers formed from thermoplastic polymers. This is the main reason why thermoplastic polymer-based textile waste is the most recycled waste [58].

To increase the quality of the recycled end product in cotton mechanical recycling, there is an obligation to use virgin fibers in addition to recycled fibers at a predetermined ratio. This can be attributed to the decrease in strength according to the recycling cycles as each cycle results in a lower degree of polymerization [59]. The upper usage limit of 30% for recycled cotton in fabrics is specified due to the shortened fibers. The amounts higher than this value causes decreases in fabric quality and performance [60]. Another study in the literature supported this result [61]. Since the fiber breakages are created in the mechanical recycling of cotton [59, 62], low-performance fabrics may be obtained not suitable for professional wear such as workwear, personal protective equipment, career wear, and uniforms [63].

Recycling is classified as primary and secondary recycling in several references. Secondary recycling can be handled as mechanical recycling and the limits mentioned above are also valid for this type of recycling. On the other hand, in primary recycling, the features of waste such as being from a single source and being pure are indisputable, while the low cycle number for each material and even the non-recyclability of some materials constitute these limits [64].

An important factor limiting the chemical recycling process of cotton is the use of harmful chemicals in the industry. While trying to minimize the damage to nature with waste disposal, the use of harmful chemicals which refers to the duality in this phenomenon creates greater harm to both nature and the consumer [59]. Moreover, the need for the separation of textile waste according to color and/or product type is inconvenient. One of the problems encountered in cotton recycling is that most of the cotton products are dyed ones and it is difficult to work with mixed-colored wastes [63]. Besides, there are studies proving that cotton fibers recycled from colored fabrics tended to possess lower quality values [65]. Thus, the demand for more environmentally friendly approaches continues [59].

Wool is a natural fiber that can only be mechanically recycled. The staple length of wool gets shorter with recycling, and it is used in blend ratios with virgin wool up to 70 recycled/30 virgin. The limited market of recycled wool is also a huge obstacle to the recycling of this fiber [63].

Nylon is a polymer with a wide variety of types that is stated as an infinitely recyclable polymer [66]. It is difficult to recycle nylon with mechanical recycling in the industry. In addition, the low number of nylon suppliers makes recycled nylon fibers more expensive [63]. Vidakis et al. studied the effects of multiple recycling cycles of PA12 on its properties. There was a decrease in mechanical properties above 5 recycling cycles. This weakening in mechanical properties is explained by the decrease in crystallinity and the beginning of degradation [67]. When the thermal recycling of polyamide 6 is evaluated, it was seen that a drying process is suggested before melting. It is revealed that the drying process prevents hydrolytic chain scission in wet materials and the intrinsic properties of PA6 polymer are remained [68].

Various studies were conducted in the literature in the last two decades for the determination of recycling cycle limits of polymers. PET which is a thermoplastic polymer widely used in the textile sector one of the polymers tested. Högg performed four recycling cycles on PET and characterized the basic polymer properties. He revealed that there was a considerable decrease in Young's modulus according to the decrease in intrinsic viscosity [69].

The polyolefin fibers react with oxygen in remelting cycles. High temperature or UV light applied in remelting cycles result in molecular weight loss. The dissolution/reprecipitation process for the recycling of high-density polyethylene (HDPE) has been suggested to overcome this limit by Poulakis and Papaspyrides. It is remarked that both the polymer and the solvent has been recovered efficiently. According to this process applied as two recycling cycles to HDPE, no changes in molecular weight, distribution of grain sizes, crystallinity, and mechanical properties were observed [70]. The same researchers applied this technique to virgin PET in pellet form and PET in blow-molded bottle form in two cycles. It was observed that the properties of recycled PET did not change [71].

The effects of seven recycling cycles on PLA (polylactic acid) polymer which is also a polyester was evaluated by Pillin et al. They observed a notable decrease in the molecular weight of the polymer. They attributed the changes in stress and strain at break, modulus, and hardness via recycling cycles to the decrease in molecular weight [72]. Another group studied eight recycling cycles of PLA and concluded that there were no changes in the mechanical properties of the polymer due to the successive cycles [73].

PAN (polyacrylonitrile) is another polymer commonly used in the production of textile products as an alternative to wool. The most critical factors limiting the recycling of PAN are the easy accessibility of perfect virgin PAN and the harsh processing conditions. From this point of view, economic conditions come into play in the recycling of PAN. The high temperature applied during recycling is also shown as a disadvantage for acrylic, which is a polymer inclines to open-loop recycling [74].

Textile wastes consisting of blends of various fibers complicate the recycling process and sometimes even make it impossible. These fibers need to be separated, which should be done by expert workers to avoid problems with the recycled final product. In addition, when it is impossible to recycle these wastes, they reach their end-of-life by utilization in energy recycling [75].

From a different point of view, there are basically two main factors limiting the recycling of textile fibers. The first one is the technological limits of recycled fibers and their inability to be used within virgin fiber, yarn, or fabric production methods. The second is that the expected product quality value cannot be reached by using these recycled fibers [76].

There are two types of sustainable fashion drivers in a sustainable fashion as production and consumption drivers. Material, human and intellectual resources form the production drivers and purchasing decisions, usage, and post usage form the latter. While technical limits are considered in the first derivatives, unawareness of consumption causes a considerable increase in waste [77]. The fact that recycled materials are generally suitable for downcycling emphasizes an important point that should be evaluated economically. Another economic point is the low consumer demand for recycled products. The reason behind the low market demand is the use of dangerous chemicals in recycled products. In addition, waste sorting is a big problem and if it is not done properly, it negatively affects the recycling process from the beginning. Finally, the relevant standards are still in their infancy. All of these may be listed as examples of the limits of recycling [78].

4.2 Branded sustainable textile fibers

Cotton and polyester are the most recycled fibers as referred before. Cotton is the most used type of natural fiber in the textile and clothing industry global consumption is reported as 26.16 million tons and the production rate is 26.43 million tons by the year 2021. When we evaluate cotton cultivation in terms of environmental aspects, it requires a large quantity usage of land occupation, water, and also pesticides. Due to pesticides, it pollutes clean water resources. In the textile production process, cotton dyeing needs a high amount of energy consumption, water, steam, and chemicals, such as bleaching agents, soap, softeners, and salts for obtaining the desired color [55, 79, 80]. Polyester is a non-biodegradable fiber in the environment. Its production process is very similar to polyamide. But polyester is extensively recycled especially as plastic bottles made of polyethylene terephthalate to reduce the landfills. Polyamide is used especially in carpets as referred before. But its recycling process is difficult because of the used dyes and chemicals added to its polymer solution [10].

Besides these types of common fibers, there have also been come out brands with the increasing recycling trend. Renewcell® technology is the upcoming brand from Sweden since 2017. For this process, used garments and textile production waste with high cellulosic content such as viscose, lyocell, modal, acetate, and other types of regenerated fibers (also called man-made cellulosic fibers) are used. Their accessories, such as buttons, zips are removed from the textile material, then it is turned into a slurry. Contaminants and non-cellulosic contents are sorted out from this slurry. This blend, brand named as Circulose® that is consisted of dissolved pulp from 100% recycled textiles dried and packaged as bales for being involved in the textile production process [81, 82].

Repreve® is known as the r-PET staple and filament yarns which are made from post-consumer water bottles and pre-consumer waste, and their fibers are used in many types of industrial product categories. Accessories, apparel, automotive, bedding, flooring, footwear, furnishings, medical accessories, military, outdoor, socks, and hosiery are some of them. As they stated they eliminated the processes; crude oil wellhead, crude oil refinery, Naptha, Xylenes, Paraxylene, TA (Terephthalic Acid) & MEG (Mono Ethylene Glycol). They have chip production (polymerization), extrusion, and texturing for Repreve® polyester filament yarns and feed stock preparation (polymerization), extrusion, and staple processing for Repreve® staple polyester fibers [83, 84]. Moreover, there is a recycled Nylon brand that is Repreve® Nylon 6 fibers. In production, they have also eliminated the processes; crude oil wellhead, crude oil refinery, benzene, cyclohexane, HMD (Hexamethylenediamine), adipic acid, and nylon salt. They have only chip production (polymerization), extrusion, and texturing processes [85].

Trevira® Sinfineco is the brand used for textiles that contain sustainable Trevira® products. They worked together with Thailand-based parent company Indorama whose manufactures recycled chips from PET bottles. They have certificates for recycled chips, fibers, and filaments from GRS (Global Recycled Standard) and RCS-NL (Recycled Claim Standard). Their products are mainly used in the automotive and apparel sectors. Trevira® Sinfineco PLA fibers and filaments are produced from plant sugars (sugar beet, sugar cane, and maize). So, they are recyclable and 100% biodegradable (industrially compostable) fiber materials. The plant sugar is subjected to the fermentation process and it is transformed into lactic acid. Besides their advantageous properties such as UV stability, fastness to light resistance, good wicking properties, it has less environmental impacts. 70% less CO_2 is emitted and 42% less energy is consumed in the raw material production process. They have ISEGA certification for PLA fiber types used in hot water filtration applications (tea and coffee filters) and packaging materials contacting with food [86].

rPET companies supply post-consumer materials in different ways. One of these interesting materials is Bionic®, which collects its source from the coastline of the oceans and waterways to produce rPET PES. Besides environmental benefits, they also get community support both for collecting and cleaning, building up waste management systems including sorting by material and color, compacting, grinding, and warehousing. Besides, they teach the system wherever their collecting point is. Then, they send them for pelletizing. Finally, the recycling process goes in the traditional way. They have three kinds of yarns; FLX® from marine plastics, DPX® from recycled plastics, and natural or synthetic fibers for gaining softer texture, HLX® from 3 layers; core, recovered with rPET and natural fibers in the outer sheath [87].

As technological sustainability process Lenzing™ introduced Refibra™ Technology which is called as reborn Tencel® Fiber Technology and they addressed that it is one of the circular economy solutions. It is a closed-loop technology in which cotton scraps and wood are used for pulping processes. For cotton scraps, they use a special and patented method for transforming colored cotton rags into the lyocell grade pulp by dye removal process and degree of polymerization adjustment. Recycling and upgrading of cotton scraps to new virgin lyocell fibers are free from water and solvent usage. It is certified according to Recycled Claim Standard (RCS) and Global Recycle Standard (GRS) [88, 89]. Lenzing™ EcoVero™ fibers are sustainable viscose fibers that are produced by the use of certified and controlled sustainable wood sources, ecological production process, and supply chain transparency as stated. It has 50% lower emissions and water impact than generic viscose. Lenzing™ EcoVero™ fibers are certified with the EU Ecolabel. It means that the production method has a lower impact on the environment compared with other products in the market [90]. Livaeco by Birla Cellulose™ is eco-enhanced viscose manufactured using a closed-loop process. As they declared, they make a series of changes in the process to be more environmental-friendly. They used a molecular tracer so that they can follow the product from fiber stage to garment form and they can verify the product easily. They emphasize that their source is from certified sustainable forests, they consumed lower water compared with other types of natural fibers, lower greenhouse gas emissions and biodegrades in 6 weeks. They stated that cost of Livaeco™ is 4–5% higher than the conventional type of their fibers produced [91]. Livaeco™ has the FSC® C135325 certificate that refers wood is sourced from the forests following the principles of Sustainable Forestry Management provided environmental, social and economic benefits. They also have various certificates, tools, and documents about sustainability for different processes. They have Forest Stewardship Council (FSC®) certificate for obtaining wood, pulping, fiber production processes regularly; Rainforest Alliance certificate in pulping process; Higg Index, Thinkstep in fiber production step; Tracer tool (fiber, yarn, fabric, garment), OEKO-TEX 100, Sustainable Textile Solution for their Livaeco™ viscose fiber, BLOCKCHAIN for Fiber 2 Retail Process. Besides these certificates, they achieved Dark Green Shirt, Ranking in Canopy's Hot Button Report in 2020 [92]. Kelheim Fibers have also CELLIANT Viscose which is introduced as the first in-fiber sustainable viscose infrared (IR) solution that is an alternative to synthetic fibers. They use natural minerals and embedded them into plant-based fibers. It is certified by FSC® or PEFC™ about raw material used. They are also awarded with a dark green/green shirt in Canopy's 2021 Hot Button Report, which is a sustainability indicator for viscose fiber producers [93].

When polyamide is considered, one of the brand marks is Econyl® by Aquafil S.P.A. It has two types of nylon textile filament yarns; ECONYL® FDY yarns on beam and ECONYL® texturized yarns on cones that are both types of yarns produced via using 100% recycled post-consumer and post-industrial recycled content. They use fishnets, carpets, oligomers (generated by polymer industries), and other types of PA6 materials as wasted content. In ECONYL® plant operation processes has two steps as depolymerization step (where the specific mix of waste is transformed back into secondary raw material-caprolactam) and the purification step of caprolactam [94].

Fulgar is another company that has various types of sustainable fibers with the brand names; Q-NOVA®, Q-CYCLE®, EVO®, AMNI SOUL ECO®. Q-NOVA® PA 6.6. yarn has an eco-friendly process called as MCS (Spinning Continuous Melting). MCS is a mechanical regeneration system that does not involve using chemical materials which would lessen the sustainability of the end product. More than half of it is produced by pre-consumption waste. This waste is remolded using a mechanical regeneration process, then after, it is turned into a form of a polymer. Its prominent features are stated as lightness, breathability, having bright colors. It has certificates as The Global Recycled Standard (GRS), EU ECOLABEL, Higg index [95]. Q-CYCLE® yarn is their new eco-sustainable PA 6.6 yarn produced with their interaction with BASF's ChemCycling™ recycling project. They use post-consumer recycled contents like plastic wastes (used tires) that is not possible to be mechanically recycled. Its certifications are under the evaluation process [96]. EVO® is the other trademark of Fulgar that is a bio-based origin polyamide that its polymerization is partially or completely sourced from castor oil (from castor seeds) [97]. AMNI SOUL ECO® has enhanced PA 6.6 formula, developed by Rhodia-Solvay group, which enables garments to be a biodegradable feature when left in landfills [98].

Considering the polyurethane known as elastane in the market, COREVA™ can be mentioned. It is a plant-based yarn obtained from natural rubber for replacing synthetic, petrol-based yarns and is patented by Candiani Denim. Organic cotton is wrapped around a natural rubber core, so they produce plastic-free yarn by replacing conventional synthetic and petrol-based elastomers. As they declared, Candiani has created an innovative, biodegradable stretch denim fabric but still, it has the features such as elasticity, physical qualities, and durability that are important factors for producing jeans [99].

4.3 Sustainability certifications for textiles and textile eco-labels

Environmental issues are trending topic and their importance increase gradually. There are some international treaties to regulate the behavior of the countries to reduce greenhouse gases and protect the ozone layer. Kyoto Protocol and Montreal Protocol are exemplary treaties for the sign of industrialized countries, describing the precautions that they should take [55]. The carbon footprint is the amount of the greenhouse gases released from fossil fuels used for electricity, heating, and transportation purposes. Textile and clothing sectors are the leading sectors that have high carbon footprint generation and greenhouse gases emissions [37]. Energy is the other critical case for the textile industry. The consumed energy according to textile processes can be given as 34% for spinning, 23% for weaving, 38% for chemical process, and 5% for various purposes [100].

All the efforts for sustainability including getting certifications, discovering new sustainable processes, producing new sustainable fibers, getting textile ecolabels, United Nations' The Sustainable Development Goals (UNSDGs) are playing a major role. United Nations' 17 goals can be listed regularly as; no poverty, zero hunger, good health and well-being, quality education, gender equality, clean water and sanitation, affordable and clean energy, decent work and economic growth, industry, innovation and infrastructure, reduced inequalities, sustainable cities and communities, responsible consumption and production, climate action, life below water, life on land, peace, justice and strong institutions, partnerships for the goals. There are various studies about the relationship between UNSDGs and fashion brands, certifications, and new type of sustainable fibers [101, 102].

In the past, products are disposed of after the end-of-life or disuse of the products. But today, solutions and precautions for sustaining the environmental cycle are steadily taken. ISO 14040:2006 (Environmental Management-Life Cycle Assessment-Principles and Framework) is the valid standard to evaluate the sustainability of the product cycle [103, 104]. Life cycle assessment (LCA) is a methodology that is determined by the ISO 14040 and ISO 14044 [80]. It merges the environmental impacts of the studied product or service through the value chain [104]. It is possible to determine the potential environmental benefits of various systems of textile reuse and recycling processes within the methods of LCA [50]. LCA does not contain design and development stages because it is considered that design of the product has not environmental impact. But the design of the product can be affected by the other life cycle stages such as emissions to air, water, and land at each stage of manufacture, use, and disposal of the product [105].

There are various textile sustainability standards and certifications. EU Ecolabel supports Europe strategy for zero pollution and circular economy targets by minimizing products' harmful impact on the environment. Products labeled with EU Ecolabel make a reduction in water consumption, make less pollution in the air, restrict the use of hazardous chemicals, and minimize the waste [106, 107]. Better Cotton Initiative (BCI) is claimed itself as the world's leading sustainability initiative for cotton. Their mission is to help cotton communities survive and thrive while protecting and restoring the environment. They have selected five impact areas consisted of climate change mitigation, soil health, pesticide use, smallholder livelihoods, and women's empowerment [108, 109]. In BCI's Better Cotton Assurance Model, they have a roadmap for Better Cotton Farmers and farmer groups to move from baseline performance to meeting the key indicators of the Better Cotton Principles and Criteria and ultimately achieving long-term improvement goals. The model has four overarching goals. The first one is giving license to sell their cotton as Better Cotton if they can meet the standards and criteria to license for selling their cotton as Better Cotton. The second one is improvement in the framework for making sustainable practices. The third one is the development in the improvement of connection between producers and partners. The last one is measuring the sustainability performance of the farmers [109, 110].

Besides OEKO-TEX Standard 100, OEKO-TEX has series of Sustainability Standards comprising of Oeko-Tex Sustainable Textile Production (STeP), Made in Green by OEKO-TEX®, ECO PASSPORT by OEKO-TEX®, OEKO-TEX® DETOX TO ZERO. ECO PASSPORT by OEKO-TEX® is used for chemical products (textile and leather chemicals, colorants, and auxiliary agents) that are used in the textile, leather, and clothing industry. Oeko-Tex Sustainable Textile & Leather Production (STeP) is the standard for modules, such as chemical management, environmental performance, environmental management, social responsibility, quality management, health, and safety in production chain. To get Made in Green by OEKO-TEX® certificate, some criteria (some OEKO-TEX® certificates) should be taken due to finished products that consumers can buy at retailers or semi-finished products sold to companies within the supply chain. This certificate means that textile or leather products' materials are tested for harmful substances, produced as environmentally, safe, and socially responsible workplaces are supplied [111].

GOTS is also one of the textile processing standard for organic fibers, also both for ecological and social criteria. It comprises the whole textile supply chain starting with harvesting of the raw materials till packing and labeling. It is important to use dyes and chemicals that have a low impact on environment and even it has water norms in production, besides this, it also considers fiber requirements, environmental criteria, social criteria, and traceability. GOTS have various production criteria limits. For example, additional fiber limits for natural fibers both for vegetable and animal fibers (linen, hemp, wool, silk, mohair, etc.) is up to 30%; for sustainable regenerated fibers is (Lyocell® & protein based fibers: from organic, FSC(Forest Stewardship Council™)/Programme for the Endorsement of Forest Certification (PEFC) certified recycled raw materials is up to 30%; for Recycled Claim Standard (RCS from Textile Exchange), Global Recycle Standard (GRS from Textile Exchange), Recycled Content Standard (from SCS) certified synthetic fibers (polyester, polyamide, polypropylene, and polyurethane) is up to 30%. There are also restricted fibers in blends like conventional cotton, virgin polyester, conventional angora hair, acrylic, asbestos, and carbon, silver. They have also an obligation for using virgin synthetic and regenerated fibers like viscose, modal, polyamide, elastane, and polypropylene in fiber blends as the maximum ratio is 10%. They have given some more examples like it is permitted to use 70% organic cotton, 30% lyocell from the organic plantation; but, it is not permitted to use 70% organic cotton, 30% lyocell from conventional wood [112, 113].

BlueSign® is one of the sustainability standards that offer a system with solutions for industry and brands for increasing their sustainability performance. They have various criteria such as chemical products for end-consumer use, surface treatment of metals, and plastics/non-textile substrates, fiber manufacturing, textile manufacturers, down and feathers processing, flame retardants, nanoscale materials/structures [114]. They have also a restricted substances list (RSL). In fiber manufacturing for production sites, it is stated that 99% solvent recovery (lyocell, acetate, etc.) rate should be aimed at dry spinning or wet spinning. They encourage their partners to develop fibers that meet their requirements for supporting a circular economy and to give ahead manufacturers to produce and use of recyclable and recycled fibers for circular textile production. It is obligatory for fiber manufacturing sites to pass the chemical assessment that they use Alkylphenol ethoxylates (APEO), free agents, in all preparation and sizing agents used. It is possible to give more examples for other type of fibers. In polyester fiber production, they have limited values of volatile organic compounds (VOCs) not only for year, but also limited emission factors per PET chips (one kg) and filament fiber (one kg). It is also important to have wood policy for cellulosic regenerated fibers, such as viscose, lyocell, and acetate. In production, 25% of sourced pulp fibers/pulp should be used from the wood certified by independent third-party certification with the label of the Forest Stewardship Council (FSC®). Besides this, independent third-party risk assessments, audits and on-site visits should be taken with positive results by audits (preferably a CanopyStyle Audit with at least bronze status) or independent third-party certification of sustainable forest management programs (e.g. Rainforest Alliance) [115].

The Higg Index is used as a tool for the standardization of sustainability measurement. It is comprised of five tools; the Higg Facility Environmental Module (FEM), Higg Facility Social & Labor Module (FSLM), Higg Brand & Retail Module (BRM), Higg Materials Sustainability Index (MSI), and Higg Product Module (PM). They evaluate the social and environmental performance of the value chain together with the environmental impacts of products. It gives an opportunity to consumers using the Higg Index to inform their individual sustainability strategies in crosswise topics, such as water use, carbon emissions, labor conditions, consumer goods brands, retailers, manufacturers, and governments [116].

The Recycled Claim Standard (RCS) and Global Recycled Standard (GRS) are stated as international and voluntary standards. They set requirements for third-party certification about recycled input and chain of custody. Their aim is to raise the usage ratios of recycled materials. The GRS contains also social and environmental processing requirements and chemical restrictions as additional criteria compared with RCS [117]. For RCS, labeling can be applied to all products containing at least 5% recycled material for textiles. It also enhances the traceability of recycled raw materials, transparent communication, clear labeling, and stakeholder engagement [118]. The GRS label assured that there are high percentages of recycled contents in products, the harmful impact is reduced both for people and the environment, traceability and stakeholder engagement are supplied [119].

Cradle to Cradle Certified® is another global standardization for safe, circular, and responsibly made products. It evaluates the safety, circularity, and responsibility of materials and products in five categories of sustainability performance such as material health, product circularity, clean air & climate protection, water and soil stewardship, and social fairness [120].

Forest Stewardship Council® (FSC) forest management certification endorsed that the management of forests is made by taking care of biological diversity and benefits the lives of local people and workers. There are 10 principles for forest operation for receiving FSC forest management certification. These principles include a broad range of issues, from maintaining high conservation values to community relations and workers' rights, as well as monitoring the environmental and social impacts of forest management [121].

There are also some other sustainability standards like Cotton Made in Africa, Organic Content Standard (OCS), Soil Association Organic Standard, Responsible Down Standard (RDS), Responsible Wool Standard (RWS) [122–126].

5. Conclusions

Recycling has shown continuity since ancient times as a technique that people comprehended its importance towards the purpose of living with scarce resources and applied it even if not in a scientific sense. Recycling has reached scientific meaning throughout history, and then the subject has evolved towards sustainability. Textile recycling has a great place within the scope of this subject, which has been on the agenda for a long time and will also continue to be, with the advantages it creates in both environmental and economic terms. Human beings fall into textile products from the moment that they are born, they need these textile products throughout their lives (even when they die in some cultures—due to the rituals of burial with various fabrics). The indispensability of textile has always kept it at the forefront in various areas for years.

Engineering-based scientific research always aims to increase the quality of life and make the world habitable for a longer period. In this context, these purposes are embodied as the main objectives in the studies on recycling and sustainability. As the decrease in natural resources, population growth, changes in fashion causing excessive consumption of resources, and technological developments continue, the interest in recycling and sustainability will increase acceleratingly. As emphasized herein, recycling in textiles, recycling limits in textile wastes, and the search for sustainable new textile resources will continue to be hot topics of the area. In conclusion, approaches on more effective utilization of traditional fibers, the discovery, commercialization, and popularization of new sustainable fibers, and the representation of new models for the management of textile waste will be the focus of researchers for years.

Conflict of interest

The authors declare no conflict of interest.

Author details

Gizem Celep[1], Gamze D. Tetik[2] and Fulya Yilmaz[3]*

1 Department of Occupational Health and Safety, Faculty of Health Sciences, Usak University, Usak, Turkey

2 Department of Materials Science and Nanotechnology Engineering, Faculty of Engineering, Usak University, Usak, Turkey

3 Department of Textile Engineering, Graduate Education Institute, Usak University, Usak, Turkey

*Address all correspondence to: fulya.ozdemir@usak.edu.tr

References

[1] Hawley JM. Textile recycling. In: Worrell E, Reuter MA, editors. Handbook of Recycling: State-of-the-art for Practitioners, Analysts, and Scientists. Burlington: Elsevier; 2014. p. 211-218

[2] Zafer Kalkınma Ajansı, Uşak İli Tekstil Geri Dönüşüm sektör Raporu [Internet]. 2019. Available from: https://www.kalkinmakutuphanesi.gov.tr/assets/upload/dosyalar/usak-tekstil-geri-donusum-raporu-tgdr.PDF [Accessed: April 23, 2022]

[3] Kleinhückelkotten S, Neitzke H. Social acceptability of more sustainable alternatives in clothing consumption. Sustainability. 2019;**11**(22):6194:1-18. DOI: 10.3390/su11226194

[4] Kasavan S, Yusoff S, Guan NC, Zaman NSK, Fakri MFR. Global trends of textile waste research from 2005 to 2020 using bibliometric analysis. Environmental Science and Pollution Research. 2021;**28**(33):44780-44794. DOI: 10.1007/s11356-021-15303-5

[5] Xie X, Hong Y, Zeng X, Dai X, Wagner M. A systematic literature review for the recycling and reuse of wasted clothing. Sustainability. 2021;**13**(13732):1-20. DOI: 10.3390/su132413732

[6] Eser B, Çelik P, Çay A, Akgümüş D. Tekstil ve konfeksiyon sektöründe sürdürülebilirlik ve geri dönüşüm olanakları. Tekstil ve Mühendis. 2016;**23**(101):43-60. DOI: 10.7216/1300759920162310105

[7] Ütebay B, Çelik P, Çay A. Textile Wastes: Status and Perspectives. In: Korlu A, editor. Waste in Textile and Leather Sectors. London, UK: Intechopen; 2020. DOI: 10.5772/intechopen.92234

[8] İstanbul Sanayi Odası Dergisi, Atık Yönetimi [Internet]. 2020. Available from: https://www.iso.org.tr/Sites/1/content/dergi/2020-09/index_60.html#page=52 [Accessed: April 23, 2022]

[9] Bhatia D, Sharma A, Malhotra U. Recycled fibers: An overview. International Journal of Fiber and Textile Research. 2014;**4**(4):77-82

[10] Chen HL, Burns LD. Environmental analysis of textile products. Clothing and Textiles Research Journal. 2006;**24**(3):248-261. DOI: 10.1177/0887302X06293065

[11] Damayanti D, Wulandari LA, Bagaskoro A, Rianjanu A, Wu HS. Possibility routes for textile recycling technology. Polymers. 2021;**13**(21):3834:1-31. DOI: 10.3390/polym13213834

[12] Rani S, Jamal Z. Recycling of textiles waste for environmental protection. International Journal of Home Science. 2018;**4**(1):164-168

[13] Labayen JPJ, Labayen IV, Yuan Q. A review on textile recycling practices and challenges. Text. 2022;**2**:174-188. DOI: 10.3390/textiles2010010

[14] Chavan RB. Environmental sustainability through textile recycling. Journal of Textile Science & Engineering. 2014;**2**:1-5. DOI: 10.4172/2165-8064.S2-007

[15] Jones AG. Motivation, mechanics and magnitude: A study of glass recycling in the Roman Empire [thesis]. Raleigh, North Carolina: North Carolina State University; 2015

[16] Varshney K. Basics of waste management and disposal practices in

the Indian context. In: Lichtfouse E, editor. Sustainable Agriculture Reviews 52. France: Springer; 2021. p. 193-222

[17] Gündüz MY. Geri Dönüşebilir Atiklarin Kontrolü ve Sifir Atik Projesi Uygulamalari: Necmettin Erbakan Üniversitesi Örneği. Necmettin Erbakan Üniversitesi: Konya; 2021

[18] Gündüzalp AA, Güven S. Waste and waste types, waste management, recycling and consumer: Çankaya municipality and instance of neighbourhood consumers. Sociological Research E- Journal. 2016;**9**:1-19

[19] Zenati DND. Recycling and cleaner production: An integrated environmental responsibility and development. Beam Journal of Economic Studies. 2022;**6**(1):590-605

[20] Byerly H, Balmford A, Ferraro PJ, Hammond Wagner C, Palchak E, Polasky S, et al. Nudging pro-environmental behavior: Evidence and opportunities. Frontiers in Ecology and the Environment. 2018;**16**:159-168. DOI: 10.1002/fee.1777

[21] Antonucci S. A Recycling Program for the New York City Garment District [Internet]. 2012. 23 p. Available from: https://commons.pratt.edu/sesresearch/wp-content/uploads/sites/157/2016/11/2012_Antonucci_S_Report.pdf [Accessed: March 01, 2022]

[22] Koch MC. Managing textile waste [thesis]. North Carolina: North Carolina State University; 1998

[23] Jain P, Gupta C. Textile recycling practices in India: A review. International Journal of Textile and Fashion Technology. 2016;**6**(6):21-35 p

[24] Hawley JM. Understanding and improving textile recycling: A systems perspective. In: Blackburn RS, editor. Sustainable Textiles: Life Cycle and Environmental Impact. Manchester, England: CRC Press; 2009. p. 179-199

[25] Domina T, Koch K. The textile waste lifecycle. Clothing and Textiles Research Journal. 1997;**15**(2):96-102

[26] Coia-Ahlman S, Groff KA. Textile wastes. Research Journal of the Water Pollution Control Federation. 1990;**62**(4):473-478

[27] Divita L, Dillard BG. Recycling textile waste: An issue of interest to sewn products manufacturers. Journal of the Textile Institute. 1999;**90**(1):14-26. DOI: 10.1080/00405009908658699

[28] Chang Y, Chen HL, Francis S. Market applications for recycled postconsumer fibers. Family and Consumer Sciences Research Journal. 1999;**27**(3):320-340

[29] Hawley JM. Textile recycling: A system perspective. In: Wang Y, editor. Recycling in Textiles. Cambridge: Woodhead Publishing; 2006. p. 7-24. ISBN-13: 978-1-85573-952-9

[30] Jamshaid H, Hussain U, Mishra R, Tichy M, Muller M. Turning textile waste into valuable yarn. Cleaner Engineering and Technology. 2021;**5**(100341):1-8. DOI: 10.1016/j.clet.2021.100341

[31] Halimi MT, Hassen MB, Azzouz B, Sakli F. Effect of cotton waste and spinning parameters on rotor yarn quality. Journal of the Textile Institute. 2007;**98**(5):437-442. DOI: 10.1080/00405000701547649

[32] Yilmaz D, Yelkovan S, Tirak Y. Comparison of the effects of different cotton fibre wastes on different yarn types. Fibres & Textiles in Eastern Europe. 2017;**4**(124):19-30

[33] Arora R, Walia A, Mehtab S. Scenario of textile waste recycling: A review. Man-Made Textiles in India. 2021;**49**(11):367-371

[34] Doba Kadem F, Özdemir Ş. Tüketici sonrası geri dönüştürülen denim kumaşların seçilmiş konfor özellikleri üzerine bir çalışma. Çukurova Üniversitesi Mühendislik-Mimarlık Fakültesi Dergisi. 2020;**35**(2):379-388

[35] Gulich B. Development of products made of reclaimed fibres. In: Wang Y, editor. Recycling in Textiles. Cambridge: Woodhead Publishing; 2006. p. 117-136. ISBN-13: 978-1-85573-952-9

[36] Vadicherla T, Saravanan D, Muthu Ram M, Suganya K. Fashion renovation via upcycling. In: Muthu SS, editor. Textiles and Clothing Sustainability: Recycled and Upcycled Textiles and Fashion. Hong Kong: Springer; 2016. p. 1-54. ISBN 978-981-10-2145-9

[37] Muthu SS, Li Y, Hu JY, Ze L. Carbon footprint reduction in the textile process chain: Recycling of textile materials. Fibers and Polymers. 2012;**13**(8):1065-1070. DOI: 10.1007/s12221-012-1065-0

[38] Tshifularo CA, Patnaik A. Recycling of plastics into textile raw materials and products. In: Nayak R, editor. Sustainable Technologies for Fashion and Textiles. United Kingdom: Woodhead Publishing; 2020. p. 311-326. ISBN: 978-0-08-102867-4

[39] Reis JMLD. Effect of textile waste on the mechanical properties of polymer concrete. Materials Research. 2009;**12**(1):63-67

[40] Payne A. Open- and closed-loop recycling of textile and apparel products. In: Muthu SS, editor. Handbook of Life Cycle Assessment (LCA) of Textiles and Clothing. Cambridge: Woodhead Publishing; 2015. p. 103-123. ISBN: 978-0-08-100169-1

[41] Hussain A, Kamboj N, Podgurski V, Antonov M, Goliandin D. Circular economy approach to recycling technologies of postconsumer textile waste in Estonia: A review. The Proceedings of the Estonian Academy of Sciences. 2021;**70**(1):80-90

[42] Filip S, Sinđelić S, Kazi Z. Textile recycling. In: Proceedings of the III International Conference Ecology of Urban Areas; 11 October 2013; Serbia. p. 173-179

[43] Yalcin-Enis I, Kucukali-Ozturk M, Sezgin H. Risks and management of textile waste. In: Gothandam K, Ranjan S, Dasgupta N, Lichtfouse E, editors. Nanoscience and Biotechnology for Environmental Applications. Environmental Chemistry for a Sustainable World. 1st ed. Cham, Switzerland: Springer; 2019. p. 29-53. DOI: 10.1007/978-3-319-97922-9_2

[44] Hawley JM. Digging for diamonds: A conceptual framework for understanding reclaimed textile products. Clothing and Textiles Research Journal. 2006;**24**(3):262-275. DOI: 10.1177/0887302X06294626

[45] Coşkun G, Başaran FN. Post-consumer textile waste minimization: A review. Journal of Strategic Research in Social Science. 2019;**5**(1):1-18

[46] Lindström K, Sjöblom T, Persson A, Kadi N. Decreasing inter-Fiber friction with lubricants for efficient mechanical recycling of textiles. In: Autex 2019 World Textile Conference; 11-15 June 2019; Ghent, Belgium.

[47] Leonas KK. The use of recycled fibers in fashion and home products. In: Muthu SS, editor. Textiles and

Clothing Sustainability: Recycled and Upcycled Textiles and Fashion. Hong Kong: Springer; 2016. p. 55-77. ISBN 978-981-10-2145-9

[48] Pensupa N. Recycling of end-of-life clothes. In: Nayak R, editor. Sustainable Technologies for Fashion and Textiles. United Kingdom: Woodhead Publishing; 2020. p. 251-309. ISBN: 978-0-08-102867-4

[49] Ribul M, Lanot A, Pisapia CT, Purnell P, McQueen-Mason SJ, Baurley S. Mechanical, chemical, biological: Moving towards closed-loop bio-based recycling in a circular economy of sustainable textiles. Journal of Cleaner Production. 2021;**326**(129325):1-13

[50] Sandin G, Peters GM. Environmental impact of textile reuse and recycling—A review. Journal of Cleaner Production. 2018;**184**:353-365

[51] Harmsen P, Scheffer M, Bos H. Textiles for circular fashion: The logic behind recycling options. Sustainability. 2021;**13**(17):9714:1-17. DOI: 10.3390/su13179714

[52] Teli MD, Valia S, Kadu K, Dahale M. Upcycling of textile materials. In: Proceeding of Global Textile Congress; 13-15 February 2015; Bangkok. p. 164-167

[53] Lin TA. Study on increasing mechanical properties of recycled multi-composition thermoplastic materials by improving interphase performance [thesis]. Nagano: Shinshu University; 2021.

[54] Zaghloul MMY, Steel K, Veidt M, Heitzmann MT. Wear behaviour of polymeric materials reinforced with man-made fibres: A comprehensive review about fibre volume fraction influence on wear performance. Journal of Reinforced Plastics and Composites. 2022;**41**(5-6). DOI: 10.1177/07316844211051733

[55] Voncina B. Recycling of textile materials. In: 7th Framework Programme MDT 'Recycling'. 2016. p. 1-37

[56] Sun X, Wang X, Sun F, Tian M, Qu L, Perry P, et al. Textile waste fiber regeneration via a green chemistry approach: A molecular strategy for sustainable fashion. Advanced Materials. 2021;**33**(48):2105174. DOI: 10.1002/adma.202105174

[57] Echeverria CA, Handoko W, Pahlevani F, Sahajwalla V. Cascading use of textile waste for the advancement of fibre reinforced composites for building applications. Journal of Cleaner Production. 2019;**208**:1524-1536. DOI: 10.1016/j.jclepro.2018.10.227

[58] Vadicherla T, Saravanan D. Textiles and apparel development using recycled and reclaimed fibers. In: Muthu SS, editor. Roadmap to Sustainable Textiles and Clothing Eco-Friendly Raw Materials, Technologies, and Processing Methods. Singapore: Springer Science Business Media; 2014. p. 139-160. DOI: 10.1007/978-981-287-065-0_5

[59] Swaby Z. Green recycling of cotton muslin: First steps of treatment and characterization [thesis]. New York: Stony Brook University; 2020

[60] Nature research Custom Media. Greener way to recycle clothes passes with flying colours [Internet]. 2022. Available from: https://www.nature.com/articles/d42473-019-00359-2 [Accessed: April 01, 2022]

[61] Johnson S, Echeverria D, Venditti R, Jameel H, Yao Y. Supply chain of waste cotton recycling and reuse: A review. AATCC Journal of Research. 2020;7(1):19-31. DOI: 10.14504/ajr.7.S1.3

[62] Asaadi S, Hummel M, Hellsten S, Härkäsalmi T, Ma Y, Michud A, et al.

Renewable high-performance fibers from the chemical recycling of cotton waste utilizing an ionic liquid. Chem Sus Chem. 2016;**9**(22):3250-3258. DOI: 10.1002/cssc.201600680

[63] Rengel A. Recycled Textile Fibres and Textile Recycling. Commissioned by the Federal Office for the Environment (FOEN). December 2017

[64] Mwanza BG. Introduction to recycling. In: Parameswaranpillai J, Rangappa SM, Rajkumar G, SiengchinS,editors.RecentDevelopments in Plastic Recycling. Singapore: Springer; 2021. p. 1-13. DOI: 10.1007/978-981-16-3627-1

[65] Ütebay B, Çelik P, Çay A. Effects of cotton textile waste properties on recycled fibre quality. Journal of Cleaner Production. 2019;**222**:29-35. DOI: 10.1016/j.jclepro.2019.03.033

[66] Islam S. Sustainable raw materials: 50 shades of sustainability. In: Nayak R, editör. Sustainable Technologies for Fashion and Textiles. United Kingdom: Woodhead Publishing; 2019. pp. 343-357. ISBN: 978-0-08-102867-4

[67] Vidakis N, Petousis M, Tzounis L, Maniadi A, Velidakis E, Mountakis N, et al. Sustainable additive manufacturing: Mechanical response of polyamide 12 over multiple recycling processes. Materials. 2021;**14**(2):466. DOI: 10.3390/ma14020466

[68] la Mantia FP, Curto D, Scaffaro R. Recycling of dry and wet polyamide 6. Journal of Applied Polymer Science. 2002;**86**(8)

[69] Högg C. PET recycling – material chemistry and performance aspects [thesis]. Avancerad Nivå: Examensarbete Inom Teknisk Kemi; 2021

[70] Poulakis JG, Papaspyrides CD. The dissolution/reprecipitation technique applied on high-density polyethylene: I. Model recycling experiments. Advances in Polymer Technology. 1995;**14**(3):237-242

[71] Poulakis JG, Papaspyrides CD. Dissolution/reprecipitation: A model process for PET bottle recycling. Journal of Applied Polymer Science. 2001;**81**(1):91-95

[72] Pillin I, Montrelay N, Bourmaud A, Grohens Y. Effect of thermo-mechanical cycles on the physico-chemical properties of poly(lactic acid). Polymer Degradation and Stability. 2008;**93**(2):321-328. DOI: 10.1016/j.polymdegradstab.2007.12.005

[73] Sikorska W, Musioł M, Zawidlak-Węgrzyńska B, Rydz J. End-of-life options for (bio)degradable polymers in the circular economy. Advances in Polymer Technology. 2021;**2021**:1-18. DOI: 10.1155/2021/6695140

[74] Adegbola TA, Agboola O, Fayomi OSI. Review of polyacrylonitrile blends and application in manufacturing technology: Recycling and environmental impact. Results in Engineering. 2020. DOI: 10.1016/j.rineng.2020.100144

[75] Piribauer B, Bartl A. Textile recycling processes, state of the art and current developments: A mini review. Waste Management & Research. 2019;**37**(2):112-119. DOI: 10.1177/0734242X18819277

[76] Chaka KT. Beneficiation of textile spinning waste: Production of nonwoven materials. Journal of Natural Fibers. 2021;**2021**:1-10. DOI: 10.1080/15440478.2021.1982437

[77] Iran S. Sustainable fashion: From production to alternative

consumption. In: Jastram S, Schneider AM, editors. Sustainable Fashion. Management for Professionals. Cham: Springer; 2018. p. 139-159. DOI: 10.1007/978-3-319-74367-7_13

[78] Leal Filho W, Ellams D, Han S, Tyler D, Boiten VJ, Paco A, et al. A review of the socio-economic advantages of textile recycling. Journal of Cleaner Production. 2019;**218**:10-20. DOI: 10.1016/j.jclepro.2019.01.210

[79] International Cotton Advisory Committee. Cotton: Review of the World Situation [Internet]. 2022. Available from: https://icac.org/Content/PublicationsPdf%20Files/4a1c8d08_b7d3_475d_8ed3_2b8179f7f9b4/Cotton-Review1-2022.pdf.pdf [Accessed: April 10, 2022]

[80] Esteve-Turrillas FA, de la Guardia M. Environmental impact of recover cotton in textile industry. Resources, Conservation and Recycling. 2017;**116**:107-115

[81] Renewcell®. About Renewcell® [Internet]. 2022. Available from: https://www.renewcell.com/en/section/about-renewcell/ [Accessed: April 10, 2022]

[82] Renewcell®. Circulose® [Internet]. 2022. Available from: https://www.renewcell.com/en/circulose/ [Accessed: April 20, 2022]

[83] Unifi®. Repreve® Filament Yarn [Internet]. 2022. Available from: https://s3.unifi.com/collateral/repreve-filament-yarn.pdf [Accessed: April 20, 2022]

[84] Unifi®. Repreve® Staple Fiber [Internet]. 2022. Available from: https://s3.unifi.com/collateral/repreve-staple-fiber.pdf [Accessed: April 10, 2022]

[85] Unifi®. Repreve® Nylon [Internet]. 2022. Available from: https://s3.unifi.com/collateral/repreve-nylon.pdf [Accessed: April 10, 2022]

[86] TREVIRA®. TREVIRA® SINFINECO, Sustainable, Innovative, High-Quality, Responsible [Internet]. 2022. Available from: https://www.trevira.de/en/trevira-brands-the-seal-of-quality-for-functional-textiles/trevira-sinfineco [Accessed: April 10, 2022]

[87] Bionic Yarn. Bionic® [Internet]. 2022. Available from: https://bionicyarn.com/our-process/ [Accessed: April 15, 2022]

[88] Weilach C. Lenzings initiative to drive circular economy in the textile World [Internet]. 2022. Available from: https://textile-zukunft.de/wp-content/uploads/2014/10/20171107_Refibra-for-GTZ-public.pdf [Accessed: April 17, 2022]

[89] Textile Exchange. Sustainability is at the core of Lenzing's strategy sCore TEN [Internet]. 2022. Available from: https://textileexchange.org/wp-content/uploads/2021/05/Lenzing-AG_Sustainability-Performance-TE-Webinar-08Jun17-3.pdf [Accessed: April 18, 2022]

[90] Carey T, Bartsch P. Lenzing™ EcoVero™. Trust in Eco-Responsible Viscose [Internet]. 2022. Available from: https://milton-factory.com/wp-content/uploads/2020/10/LENZING-ECOVERO.pdf [Accessed: April 19, 2022]

[91] Livaeco™ by Birla Cellulose. Why Livaeco by Birla Cellulose? [Internet]. 2022. Available from: https://www.livaeco.com/ [Accessed: April 20, 2022]

[92] Livaeco™ by Birla Cellulose. Introducing natural fluid fashion now eco enhanced, Livaeco™ by Birla Cellulose [Internet]. 2022. Available from: https://www.birlacellulose.com/download_files/

Livaeco_pdf_1614937009.pdf [Accessed: April 21, 2022]

[93] Kelheim Fibres. Celliant Viscose, Engineered By Kelheim Fibres. Nature-based Fibers. Infrared Wellness [Internet]. 2022. Available from: https://celliant.com/celliant-viscose/ [Accessed: April 22, 2022]

[94] Aquafil Global. Environmental Product Declaration for ECONYL®, Nylon Textile Filament Yarns [Internet]. 2022. Available from: https://www.aquafil.com/assets/uploads/EPD-FOR-ECONYL-YARN.pdf [Accessed: April 10, 2022]

[95] FULGAR®. Q-Nova® [Internet]. 2022. Available from: https://www.fulgar.com/eng/products/q-nova [Accessed: April 09, 2022]

[96] FULGAR®. Q-Cycle® [Internet]. 2022. Available from: https://www.fulgar.com/eng/products/Q-CYCLE [Accessed: April 12, 2022]

[97] FULGAR®. EVO® [Internet]. 2022. Available from: https://www.fulgar.com/eng/products/evo [Accessed: April 13, 2022]

[98] FULGAR®. AMNI SOUL ECO® [Internet]. 2022. Available from: https://www.fulgar.com/eng/products/amni-soul-eco [Accessed: April 01, 2022]

[99] Candiani Store. Coreva™ [Internet]. 2022. Available from: https://candianidenim.store/en/coreva/ [Accessed: April 12, 2022]

[100] Aneja AP. Textile recycling: An environmental perspective. In: XIIIth International Izmir Textile and Apparel Symposium; April 2-5, 2014; Turkey

[101] UNDP. What are the Sustainable Development Goals? United Nations [Internet]. 2022. Available from: https://www.undp.org/sustainable-development-goals? [Accessed: April 12, 2022]

[102] Bartl A, Hackl A, Mihalyi B, Wistuba M, Marini I. Recycling of fibre materials. Process Safety and Environmental Protection. 2005;**83**(4):351-358

[103] ISO 14040:2006 (Environmental management-Life Cycle Assessment-Principles and Framework) [Internet]. 2022. Available from: https://www.iso.org/standard/37456.html [Accessed: April 10, 2022]

[104] Dahlbo H, Aalto K, Eskelinen H, Salmenperä H. Increasing textile circulation-consequences and requirements. Sustainable Production and Consumption. 2017;**9**:44-57

[105] Woolridge AC, Ward GD, Phillips PS, Collins M, Gandy S. Life cycle assessment for reuse/recycling of donated waste textiles compared to use of virgin material: An UK energy saving perspective. Resources, Conservation and Recycling. 2006;**46**(1):94-103

[106] European Commision. EU Ecolabel [Internet]. 2022. Available from: https://ec.europa.eu/environment/ecolabel/documents/EU_Ecolabel_Flyer_v3_WEB.pdf [Accessed: April 10, 2022]

[107] European Commision. EU Ecolabel Clothing and textiles/Textile products [Internet]. 2022. Available from: http://ec.europa.eu/ecat/category/en/14/textile-products [Accessed: April 10, 2022]

[108] Better Cotton™. Who we are? Better Cotton Initiative [Internet]. 2022. Available from: https://bettercotton.org/tr/who-we-are/ [Accessed: April 01, 2022]

[109] Better Cotton™. 2030 Strategy: What you need to know. Better

Cotton™ [Internet]. 2022. Available from: https://bettercotton.org/wp-content/uploads/2021/12/Better-Cotton-2030-Strategy-One-Page-Overview_10Dec2021.pdf [Accessed: April 05, 2022]

[110] Better Cotton™. Demonstrating Compliance and Continual Improvement: Assurance Program. Better Cotton™ [Internet].2022. Available from: https://bettercotton.org/tr/what-we-do/demonstrating-compliance-assurance-programme/ [Accessed: April 05, 2022]

[111] OEKO-TEX®. OEKO-TEX Standards [Internet]. 2022. Available from: https://www.oeko-tex.com/en/downloads?tx_avsite_downloadoverview%5Baction%5D=filter&tx_avsite_downloadoverview%5Bcontroller%5D=Download&cHash=d88c0d48731db4655077cf181b159ffd [Accessed: April 12, 2022]

[112] Sumit Gupta (India Representative, GOTS). GOTS, an ideal tool for sustainable textiles [Internet]. 2022. Available from: https://global-standard.org/images/stories/Sumit_Interview_in_ITJ-_Sep_2014.pdf [Accessed: April 12, 2022]

[113] Sumit Gupta (India Representative, GOTS). GOTS Criteria and Requirements Composition for Additional Fibres & Accessories [Internet]. 2022. Available from: https://global-standard.org/images/events/GOTS_Criteria_and_Requirements_Additional_Fibres_used_in_GOTS_Products_2.pdf [Accessed: April 12, 2022]

[114] Bluesign®. Blue Way [Internet]. 2022. Available from: https://www.bluesign.com/en [Accessed: April 12, 2022]

[115] Bluesign®. Bluesign® CRITERIA for production sites ANNEX: Fiber manufacturing Version 1.0/2020-03 [Internet]. 2022. Available from: https://www.bluesign.com/downloads/criteria-2020/annex_fiber_manufacturing_2020-03.pdf [Accessed: April 12, 2022]

[116] Sustainable Apparel Coalition. The Higg Index [Internet]. 2022. Available from: https://apparelcoalition.org/the-higg-index/ [Accessed: April 13, 2022]

[117] Textile Exchange. Recycled Claim Standard (RCS) + Global Recycled Standard (GRS) [Internet]. 2022. Available from: https://textileexchange.org/standards/recycled-claim-standard-global-recycled-standard/?gclid=CjwKCAjwsJ6TBhAIEiwAfl4TWMzPxBPWtCrZSEjfZ5o0ZHoEr7TkZCM8BcrCjDTBN74QxiZgWm44oRoCE1gQAvD_BwE [Accessed: April 14, 2022]

[118] Group ECOCERT. Recycled textiles, Recycled Claim Standard (RCS) [Internet]. 2022. Available from: https://www.ecocert.com/en-US/certification-detail/Recycled-textiles-rcs [Accessed: April 14, 2022]

[119] Group ECOCERT. Recycled textiles, Global Recycled Standard (GRS) [Internet]. 2022. Available from: https://www.ecocert.com/en-US/certification-detail/Recycled-textiles-grs [Accessed: April 14, 2022-04-14]

[120] Cradle to Cradle Products Innovation Institute Inc. What is Cradle to Cradle Certified®? [Internet]. 2022. Available from: https://www.c2ccertified.org/get-certified/product-certification [Accessed: April 18, 2022]

[121] Forest Stewardship Council®. Forest Management Certification [Internet]. 2022. Available from: https://fsc.org/en/forest-management-certification [Accessed: April 18, 2022]

[122] Otto Group. Cotton made in Africa [Internet]. 2022. Available from:

https://cottonmadeinafrica.org/en/about-us/ [Accessed: April 17, 2022]

[123] Textile Exchange. Organic Content Standard (OCS) [Internet]. 2022. Available from: https://textileexchange.org/standards/organic-content-standard/ [Accessed: April 17, 2022]

[124] Soil Association. Types of organic textile certification [Internet]. 2022. Available from: https://www.soilassociation.org/certification/fashion-textiles/types-of-certification/ [Accessed: April 17, 2022]

[125] Textile Exchange. Responsible Down Standard (RDS) [Internet]. 2022. Available from: https://textileexchange.org/standards/responsible-down/ [Accessed: April 11, 2022]

[126] Responsible Wool Standard (RWS) [Internet]. 2022. Available from: https://textileexchange.org/responsible-wool/ [Accessed: April 11, 2022]

Section 3

Textile Investigation

Chapter 6

Numerical Investigation of Braided Structure Potential as a Cast for Femur Shaft Fracture

Jerry Ochola and Michele Conti

Abstract

Femur fractures are repaired using orthopedic implants involving external and internal fixators. Meanwhile, tubular braided structures have not been considered for bone-shaft fracture repair, despite their potential in use as orthopedic casts. This study investigates potential of using bi-axial braided structures as casts for femur diaphysis fracture under bending loads. The braided structure model was developed using a python script while a hollow femur bone shaft was created in a 3D interface using FE Analysis methods in ABAQUS (v17) from a femur bone model rendered using MIMICS from femur bone CT scan. Numerical methods were used to investigate the change in bone shape eccentricity due to bending loads in-terms of load carrying capacity, bone curvature, bending resistance and stresses in the bone shaft. The results portrayed influence of the braided structure in ensuring the stresses due to the bending load are distributed evenly on the femur shaft surface.

Keywords: femur, braid, biaxial, python, crack, diaphysis, FEM

1. Introduction

Human Femur bone is a natural composite material consisting of a cellular component and an extra cellular component. It consists of the cortical bone and the trabecular bone [1]. The femur bone's cortical bone is the primary load carrying material. This is attributed to the presence of osteon density and fraction of osteonal structures within the cortical bone layer of the femur structure [2]. It has also been shown that tissue strength in the cortical region is superior compared to other sections of the femur bone [3]. In the course of femur use during motion the cortical bone is loaded mostly by bending moments, resulting in a high percentage of tensile strain. Even though cortical shell and the trabecular matrix contribute substantially to the strength of the peripheral skeleton [4], the cortical bone has also been attributed with supporting most of the loads on the human skeleton. The toughness of femur bone also known as energy to failure, which is the bones property associated to its capability to absorb energy during failure has been linked to the risk of fracture in the bones. Nevertheless, even though the bone structure is susceptible to rupture, there is evidence that the living tissue material within its micro-structure is capable of self-repair [5]. This special property can be exploited during repair of femur bones.

Conventionally, fracture in femur's cortical bone has been done using metal plates [6]. This repair technique using plates is normally associated with stress shielding that may lead to resorption and osteoporosis [7] due to insufficient physiological loading on the bone [8]. which eventually can cause bone refractures when the plate is removed especially in fracture in diaphyseal region [9].

An attempt of replacing the conventional bone repair techniques have been done by using braided [8, 9] and knitted [10] composite casts to repair bone fractures. The results from the use of the braided structures were associated with reduction in stress concentrations in the fractured ends of the bone than a plate. It was also reported that The tubular cast has been shown to be a promising fixation method for healing broken bones [8]. This could be linked to the fact that braided structures used as bone casts have demonstrated better ability for distributing stress [11].

The use of Finite Element (FE) modeling has been effective in the analysis of bone mechanics [12]. Further, FE approaches have been used in predicting modes of bone failure under stance and fall configurations [13] and also in the analysis of the composite reinforcement of femur bone using composite casts [8].

In this paper, we propose a braided structure with enough rigidity to reinforce the bone at fracture site. The braided structure potentially offers support and enough stiffness to the bone around the site of fracture. This study applies numerical and analytical techniques to demonstrate feasibility of using tubular braided structures in reinforcing fractured femur bones under single stance conditions.

2. Design of braided reinforcement structure

The reinforcement consists of a braided structure and fixators at both ends of the structure. The design of the braided structure was done by modeling a circular coordinate of a helical path in a three-dimensional space in both clockwise and counter-clockwise directions. The general coordinates of the helical yarn path in the clockwise and anticlockwise direction as (**Figure 1**):

Clockwise direction:

$$X_i = [r_o + r(\theta)] \cos[-\theta - (i-1)\beta], \quad i = 1, 2, \cdots n \tag{1}$$

$$Y_i = [r_o + r(\theta)] \sin[-\theta - (i-1)\beta], \quad i = 1, 2, \cdots n \tag{2}$$

$$Z_i = r_o \tan\alpha\theta, \quad i = 1, 2, \cdots n \tag{3}$$

where:

$$r_o = \frac{2p\cos\alpha}{\beta} \tag{4}$$

$$r(\theta) = \frac{a}{2} \sin\left(\frac{2\pi}{\beta}\theta + \frac{3\pi}{2}\right) \tag{5}$$

Anti-clockwise direction:

$$X_i = [r_o + r(\theta)] \cos[\theta + (i-1)\beta], \quad i = 1, 2, \cdots n \tag{6}$$

$$Y_i = [r_o + r(\theta)] \sin[\theta + (i-1)\beta], \quad i = 1, 2, \cdots n \tag{7}$$

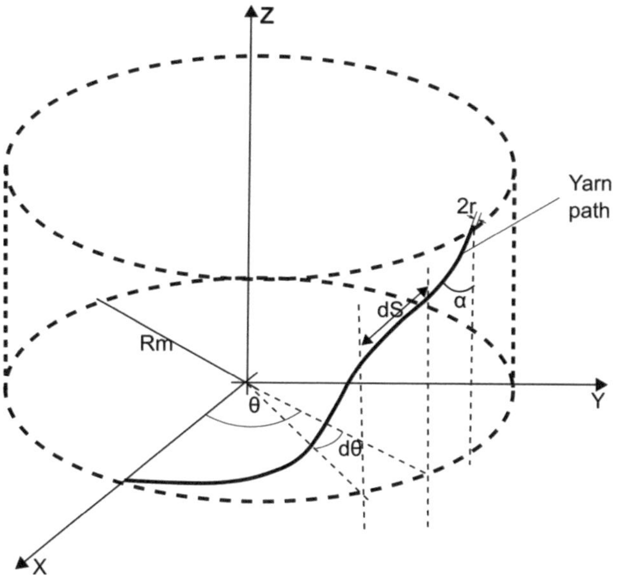

Figure 1.
Illustration of the geometry of the yarn path on a tubular braided structure.

$$Z_i = r_o \tan\alpha\theta, \quad i = 1, 2, \cdots n \tag{8}$$

In which the values of r_o are the same as Eq. (4) and $r(\theta)$ can be evaluated as: where:

$$r_o = \frac{2p\cos\propto}{\beta} \tag{9}$$

$$r(\theta) = \frac{a}{2}\sin\left(\frac{2\pi}{\beta}\theta + \frac{\pi}{2}\right) \tag{10}$$

The generic designs of braided fabrics can be modeled by formulating the coordinates for the yarn path by imputing the values of the parameters r_o and $r(\theta)$ into Eqs. (6)–(8) to (9)–(11). In the case of a diamond braided fabric, the parameters can be evaluated as follows:

Clockwise direction:

$$r_o = \frac{(2p + p')\cos\propto}{\beta} \tag{11}$$

$$r(\theta) = \begin{cases} \frac{-a}{2} & 0 < \theta < \frac{k_1\beta}{2}, \\ \frac{a}{2\sin\left(\frac{2\pi}{\beta}\theta + \frac{3\pi}{2}\right)} & \frac{k_1\beta}{2} < \theta < \frac{(k_1+1)\beta}{2}, \\ \frac{-a}{2} & 0 < \theta < \frac{k_1\beta}{2}, \\ \frac{a}{2\sin\left(\frac{2\pi}{\beta}\theta + \frac{3\pi}{2}\right)} & \frac{k_1\beta}{2} < \theta < \frac{(k_1+1)\beta}{2}, \end{cases} \tag{12}$$

Anti-clockwise direction:
the value for r_o can be evaluated using Eq. (11), then,

$$r(\theta) = \begin{cases} \dfrac{-a}{2} & 0 < \theta < \dfrac{k_1\beta}{2}, \\ \dfrac{a}{2\sin\left(\dfrac{2\pi}{\beta}\theta + \dfrac{3\pi}{2}\right)} & \dfrac{k_1\beta}{2} < \theta < \dfrac{(k_1+1)\beta}{2}, \\ \dfrac{-a}{2} & 0 < \theta < \dfrac{k_1\beta}{2}, \\ \dfrac{a}{2\sin\left(\dfrac{2\pi}{\beta}\theta + \dfrac{3\pi}{2}\right)} & \dfrac{k_1\beta}{2} < \theta < \dfrac{(k_1+1)\beta}{2}, \end{cases} \tag{13}$$

3. Materials and methods

The study involves 3D CAD developed human femur bone. Quasi-Static structural analysis was carried out using ABAQUS17 to determine the load, stress, and deformation criterion of the bone with fracture before, during and after repair with a braided structure. The model of the braided structure was developed using a python script using the parameters illustrated in **Figure 2**.

A force of 800 N was applied to induce bone displacement to simulate two configurations of single stance conditions of a human femur: (SC1, θ = 120 and SC2, θ = 90) as shown in **Figure 3**.

The profile of the femur mid-shaft was isolated inform of a surface using the commercial software paraview and then the center-line of the shaft surface traced in VMTK after-which a MATLAB algorithm was used to develop a crimper model for deploying the circular braided structure onto the oblique cut-femur bone as shown in **Figure 4** in ABAQUS CAE.

The 3D model of the braided structure was developed using a python script, an input file generated and imported into ABAQUS17. The assembly of the bone and braided structure and quasi-static structural analysis was carried out in ABAQUS17. The material properties assigned for the cortical bone are as shown in **Table 1**. The

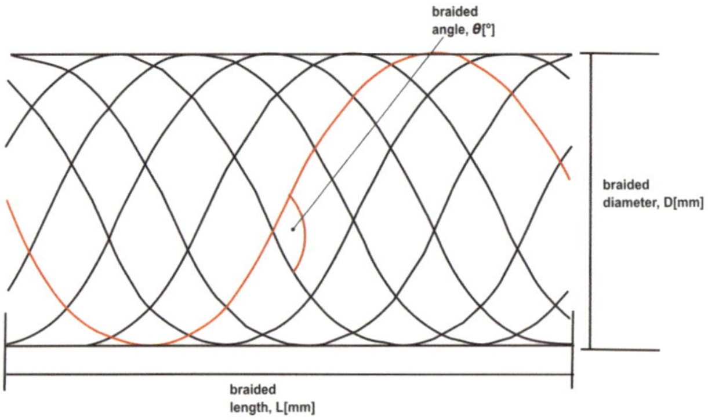

Figure 2.
Illustration of the geometry of braided structure.

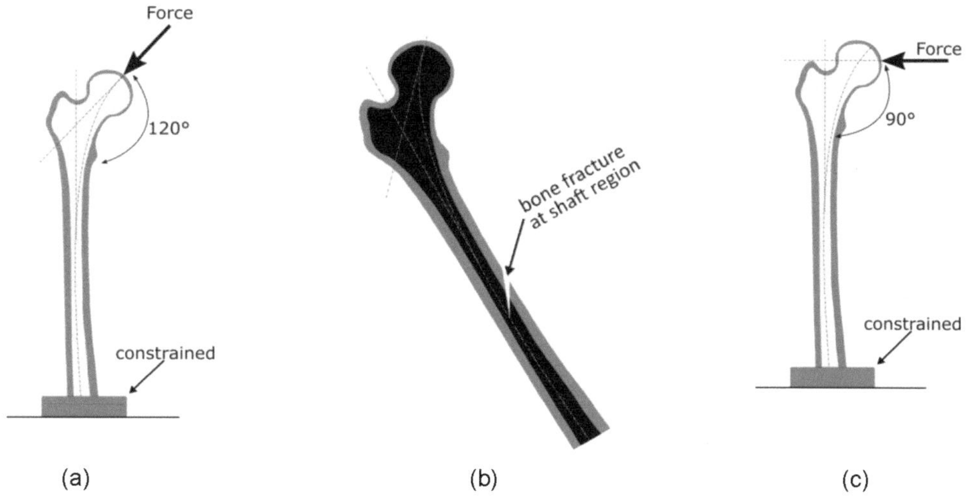

Figure 3.
Models of the femur bone structure showing: (a) single stance configuration (SC1 θ = 120); (b) position of the femur fracture at the bone shaft; and (c) ingle stance configuration (SC2 θ = 90).

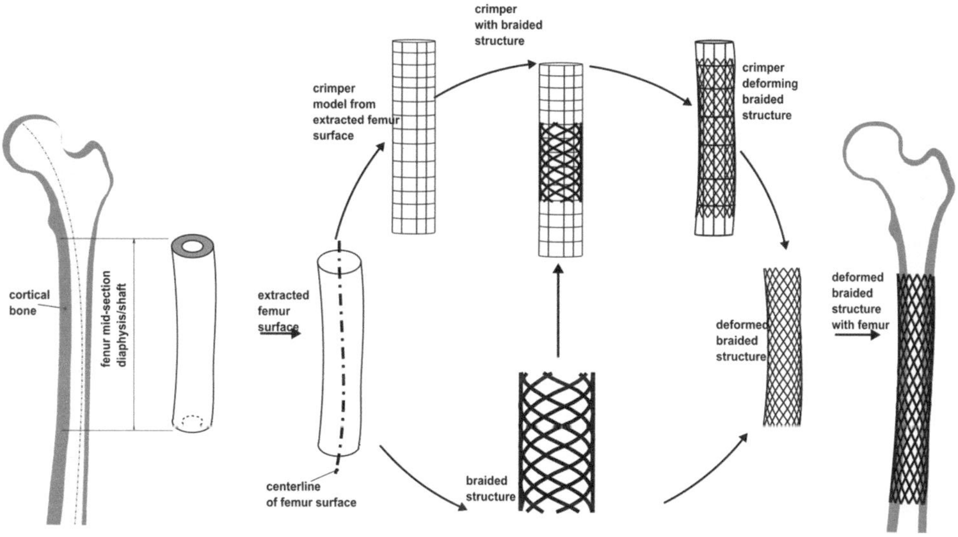

Figure 4.
Illustration of the numerical models for deployment of the braided structure onto the cut-femur model.

femur bone was modeled as a linear elastic material to reduce computation time and complexity of the analysis [15].

Quasi-Static structural analysis was carried out for both the models of the single stance configurations (SC1, θ = 120 and SC2, θ = 90). The lower end of the femur bone was fixed to mimic the normal human stance condition in all the analysis. Force applied was 800 N and was applied on femoral head at 120 and 90. The total load in the femur von Misses stress and total deformation in Z axis were evaluated for an intact femur, then for a fracture femur and eventually for a reinforced femur.

SI units	Material properties
Young's modulus	17 GPa
Density	2 gcm^{-3}
Poisson's ratio	0.30
Tensile strength	130 MPa

Table 1.
Material properties of cortical bone [14].

4. Validation

The Finite Element (FE) models adopted in this study were validated using data from previous research [16] as shown in **Figure 5**. The data shows close correlation between our model data and experimental data of femur bone analysis. The data was then used in the analysis of fall and stance configuration for the femur bone used in this study.

5. Results and discussions

The results from the numerical simulation of a single stance configurations (SC1 and SC2) for a human intact femur bone were plotted as shown in **Figure 6a** for load against displacement. It was established from the results that in an intact femur the load due to SC1 will be more than that in a SC2 type of stance configuration. The results further show that the intact bone during SC1 stance could withstand more stress than in the SC2 stance as shown in **Figure 6b**.

A fracture was then introduced to the intact femur. The force of 800 N was then applied to simulate a single stance configuration. The results of numerical simulation were plotted as shown in **Figure 7a** for load against displacement. It was established from the results that in a fractured femur the load due to SC1 stance will be more than that in an SC2 configuration. The results further show that the fractured bone during SC1 stance could withstand more stress than in the SC2 configuration as shown in **Figure 7b**.

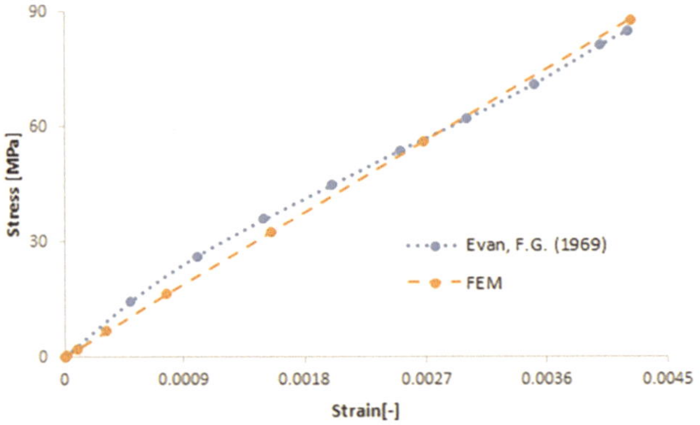

Figure 5.
Validation results for 3D femur model.

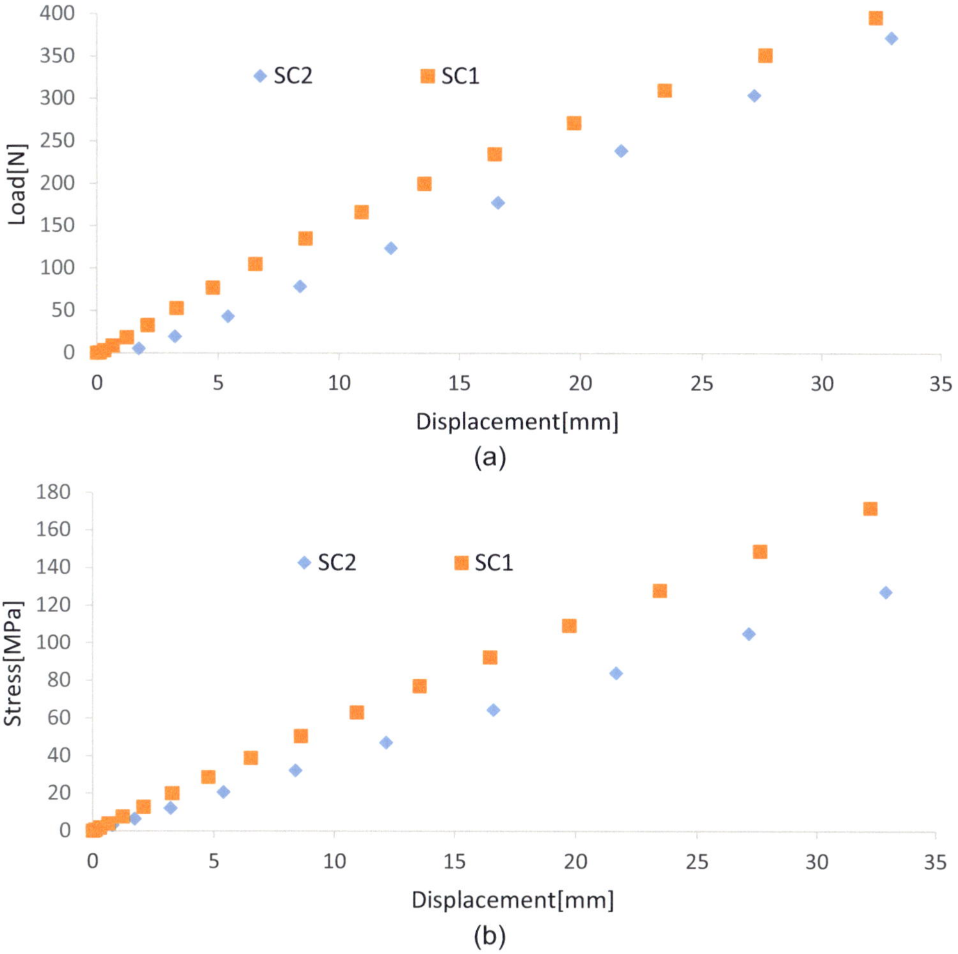

Figure 6.
(a) The illustration of load against displacement for human femur fall and single stance configuration for an intact bone structure; and (b) the illustration of stress against displacement for human femur fall and single stance configuration for an intact bone structure.

The model of the fractured femur was then reinforced using a tubular braided structure. To simulate fixators on the braided structures tie-constraints in ABAQUS were introduced at the ends of the the braided structure. An 800 N was then applied to simulate a single stance and fall configurations. The results of numerical simulation were plotted as shown in **Figure 8a** for load against displacement. It was established from the results that in the reinforced femur the load due to SC1 stance was more than that in an SC2 configuration. The results further show that the reinforced bone during SC1 could withstand more stress than in the SC2 configuration as shown in **Figure 8b**.

The deformation of the femur bone was analyzed for the SC1 conditions because it had higher mechanical properties than the SC2 stance as shown in the results. The deformation results established as shown in **Figure 9** that in an intact femur there was an increase in bone deformation with yielding stress. This was illustrated by the contour plots for the von Misses stress on the surface of the femur bone.

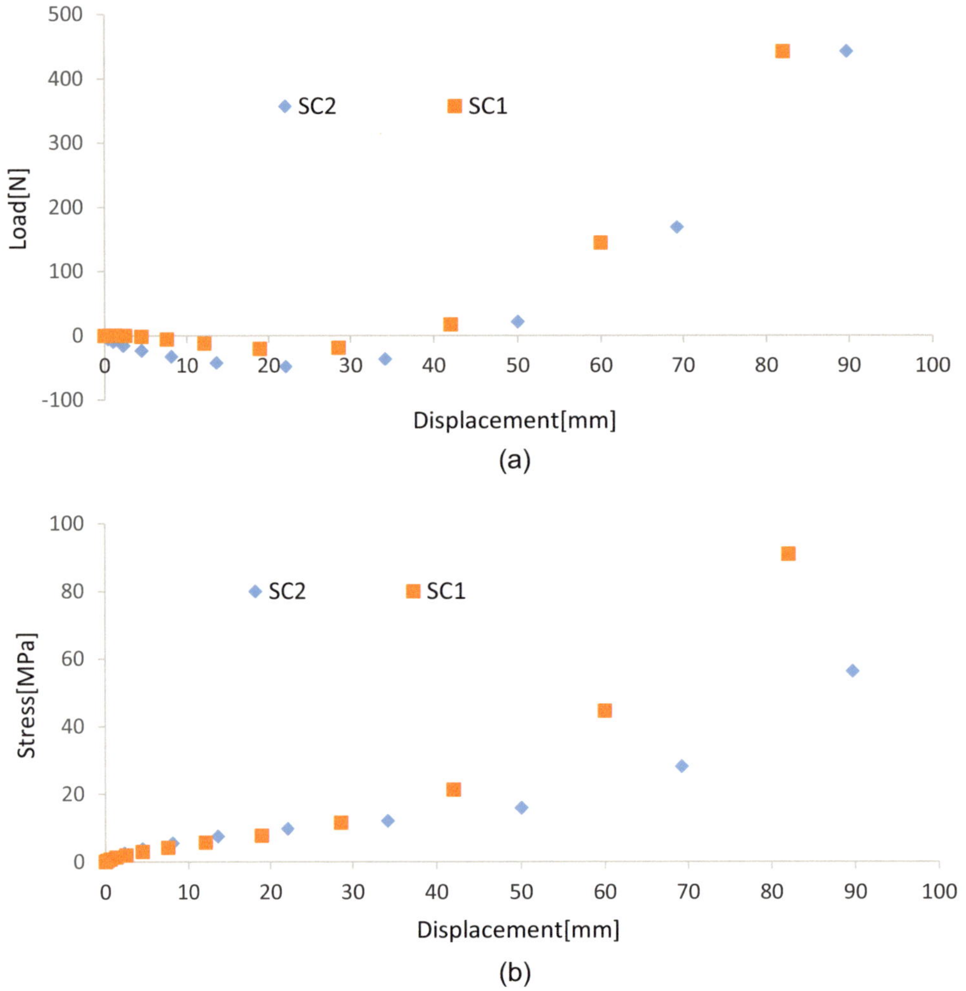

Figure 7.
(a) The illustration of load against displacement for human femur fall and single stance configuration for a fractured bone structure; and (b) the illustration of stress against displacement for human femur fall and single stance configuration for a fractured bone structure.

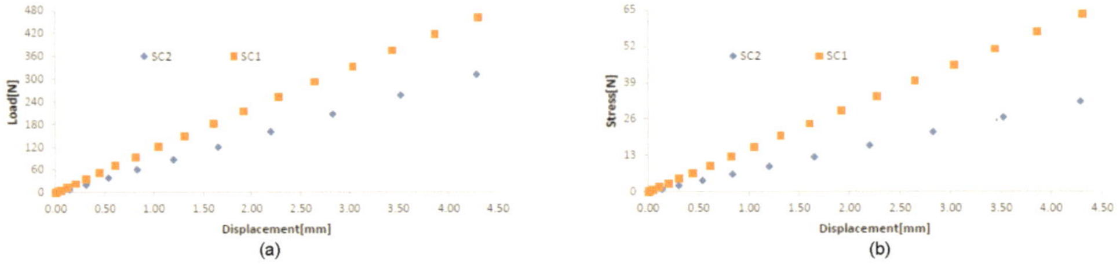

Figure 8.
(a) The illustration of load against displacement for human femur fall and single stance configuration for a braid-reinforced femur bone structure; and (b) the illustration of stress against displacement for human femur fall and single stance configuration for a braid-reinforced femur bone structure.

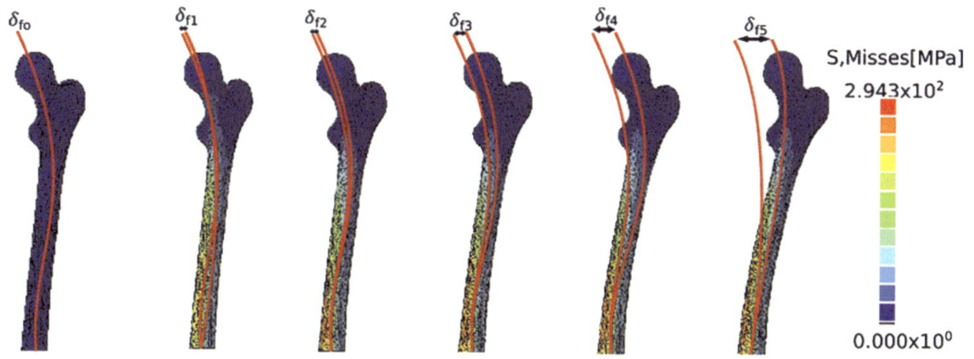

Figure 9.
Contour plots for the deformation of the model of the intact femur bone under stance configuration.

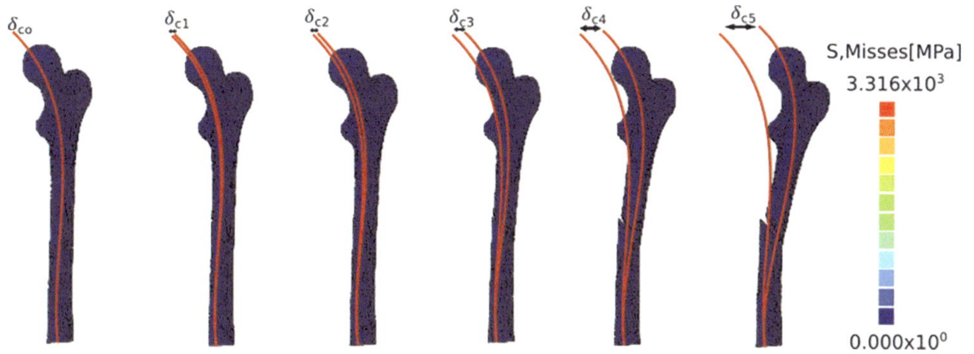

Figure 10.
Contour plots for the deformation of the model of the fractured femur bone under stance configuration.

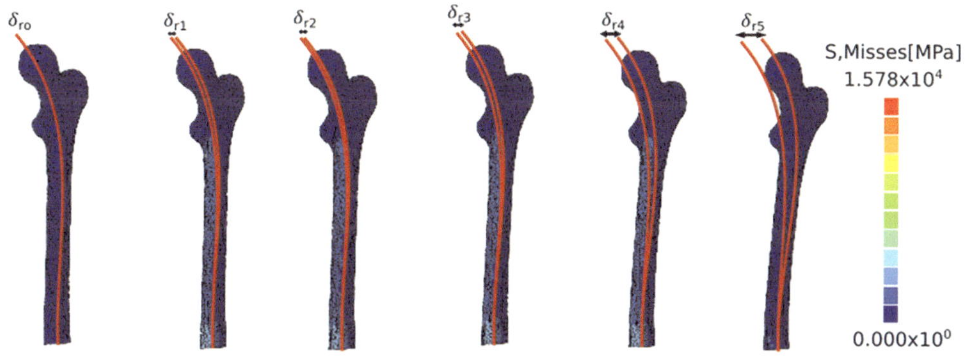

Figure 11.
Contour plots for the deformation of the model of the braid-reinforced femur bone under stance configuration.

The deformation in the fractured femur shown in **Figure 10** shows that the femur bone deformation would be higher than that of the intact femur, there was also evidence of the bone yielding at lower stress levels portrayed by the levels of von Misses stress contour plots.

The reinforced femur structure (**Figure 11**) however, illustrated lower deformations as compared to both the intact femur model and the fractured femur model.

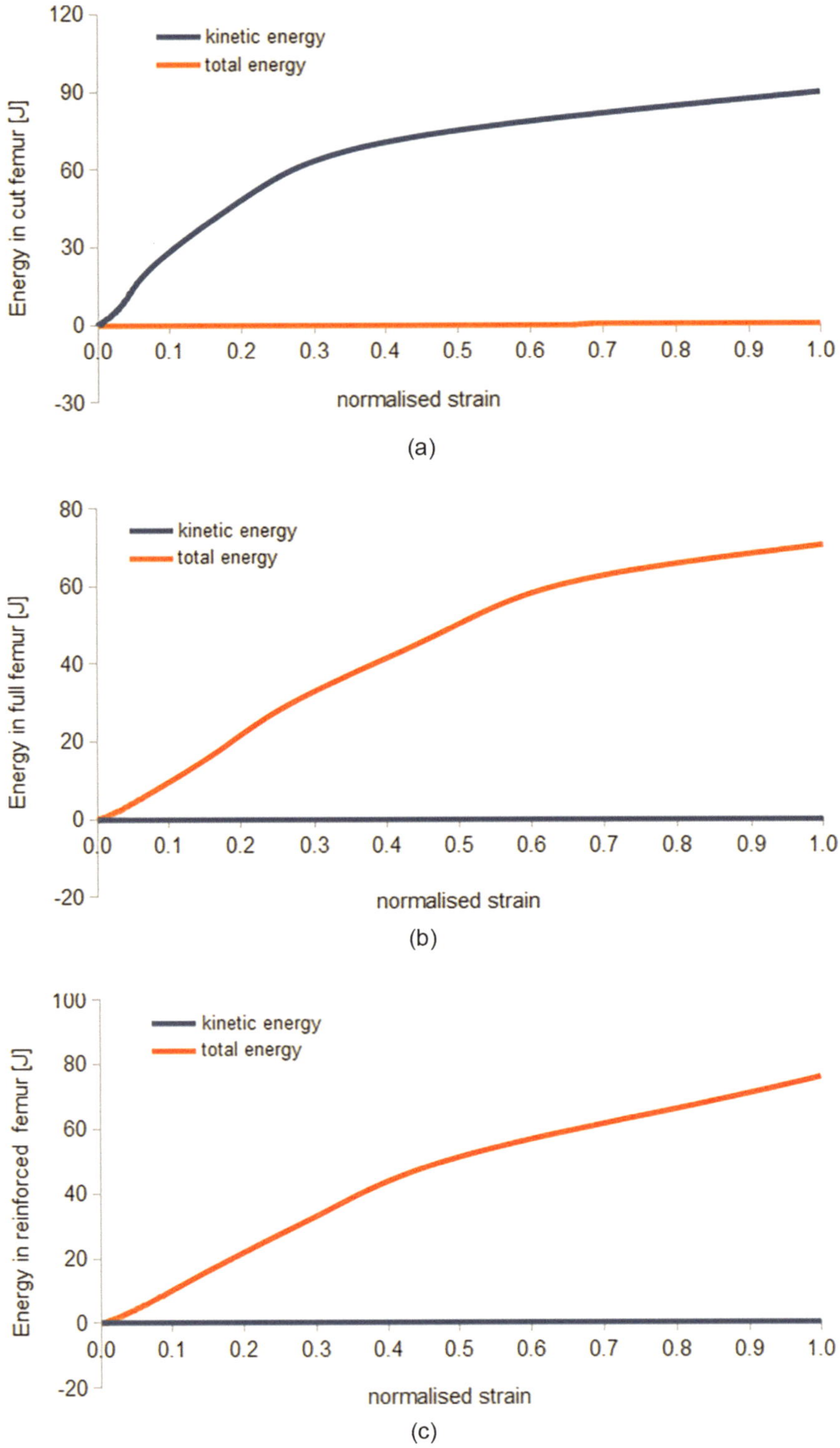

Figure 12.
The illustration of quasi-static energy analysis for the simulations for (a) intact femur bone structure; (b) fractured femur bone structure; and (c) reinforced human femur bone structure.

The results predicted in simulation models shows a trend where the load values were higher than the values in the fall configuration, these results were consistent with the findings elsewhere [17] where the fracture load in a stance configuration recorded higher loads than in a fall configuration of the femur bone. The results further established that even though the force applied to the femur was able to deform the bone to some extent, the reinforced bone structure was able to withstand the loads better. This was illustrated by the small deflection and minimum yield stress in the reinforced femur. This is further supported by the large deflection in the femur bone with the obliquecut, when the cut was reinforced using the braided structure the deflection decreased.

6. Quasi-static analysis

The simulation results shows that the models were not affected by viscous energies as portrayed in **Figure 12**.

7. Conclusion

The reinforced fractured femur portrayed improved strength and stability which could be attributed to the braided structure. There was also evidence from deformation results that showed that when reinforced the fractured femur, deformed less. Further, the use of finite element methods was found to be appropriate in the study of the feasibility of reinforcing fractured femur with braided structure.

Acknowledgements

The support from Stellenbosch Institute for Advanced Study (STIAS), University of Stellenbosch is appreciated.

Author details

Jerry Ochola[1,2]* and Michele Conti[3]†

1 Department of Manufacturing, Industrial and Textile Engineering, Moi University, Eldoret, Kenya

2 Stellenbosch Institute for Advanced Study (STIAS), Wallenberg Research Centre at Stellenbosch University, Stellenbosch, South Africa

3 Department of Civil Engineering and Architecture—Structural Division, University of Pavia, Pavia, Italy

*Address all correspondence to: ochollas@gmail.com

† This author contributed equally.

References

[1] Vaz MF, Canhao H, Fonseca JE. Bone: A Composite Natural Material, Advances in Composite Materials-Analysis of Natural and Man-Made Materials, Pavla Tesinova. London, UK: IntechOpen; 2011. DOI: 10.5772/17523. Available from: https://www.intechopen.com/books/advances-in-composite-materials-analysis-of-natural-and-man-made-materials/bone-a-composite-natural-material

[2] Wachter NJ, Krischak GD, Mentzel M, Sarkar MR, Ebinger T, Kinzl L, et al. Correlation of bone mineral density with strength and microstructural parameters of cortical bone in vitro. Bone. 2002; **31**(1):90-95. DOI: 10.1016/S8756-3282(02)00779-2

[3] Bayraktar HH, Morgan EF, Niebur GL, Morris GE, Wong EK, Keaveny TM. Comparison of the elastic and yield properties of human femoral trabecular and cortical bone tissue. Journal of Biomechanics. 2004;**37**(1):27-35. DOI: 10. 1016/S0021-9290(03)00257-4

[4] Augat P, Reeb H, Claes LE. Prediction of fracture load at different skeletal sites by geometric properties of the cortical shell. Journal of Bone and Mineral Research. 2009;**11**(9):1356-1363. DOI: 10.1002/jbmr.5650110921

[5] Augat P, Schorlemmer S. The role of cortical bone and its microstructure in bone strength. Age and Ageing. 2006; **35**(S2):ii27-ii31. DOI: 10.1093/ageing/afl081

[6] Deshmukh RM, Kulkarni SS. A review on biomaterials in orthopedic bone plate application. International Journal of Current Engineering and Technology. 2015;**5**(4):2587-2591

[7] Akeson WH, Woo SLY, Coutts RD, Matthews JV, Gonsalves M, Amiel D. Quantitative histological evaluation of early fracture healing of cortical bones immobilized by stainless steel and composite plates. Calcified Tissue Research. 1975;**19**(1):27-37. DOI: 10.1007/bf02563988

[8] Evans KR, Carey JP. Feasibility of a braided composite for orthopedic bone cast. The Open Biomedical Engineering Journal. 2013;**7**(1):9-17. DOI: 10.2174/1874120701307010009

[9] Houle JB, Tabrizi P, Giachino AA, Benoit MY, Richards RS, Pham B, et al. Refracture rate after plate removal from the radial metaphysis. Canadian Journal of Surgery. 2002;**45**(1):53-56

[10] Fujihara K, Huang Z-M, Ramakrishna S, Satknanantham K, Hamada H. Feasibility of knitted carbon/PEEK composites for orthopedic bone plates. Biomaterials. 2004;**25**(17):3877-3885. DOI: 10.1016/j.biomaterials.2003.10.050

[11] Nasiri F, Ajeli S, Semnani D, Jahanshahi M, Emadi R. Design, fabrication and structural optimization of tubular carbon/Kevlar/PMMA/graphene nanoplate composite for bone fixation prosthesis. Biomedical Materials. 2018;**13**(4):045010. DOI: 10.1088/1748-605x/aab8d6

[12] Parashar SK, Sharma JK. A review on application of finite element modelling in bone biomechanics. Perspectives in Science, Recent Trends in Engineering and Material Sciences. 2016;**8**:696-698

[13] Schileo E, Balistreri L, Grassi L, Cristofolini L, Taddei F. To what extent can linear finite element models of human femora predict failure under stance and fall loading configurations? Journal of Biomechanics. 2014;**47**(14):

3531-3538. DOI: 10.1016/j.jbiomech.2014.08.024

[14] Sabatini AL, Goswami T. Hip implants VII: Finite element analysis and optimization of cross-sections. Materials & Design. 2008;**29**(7):1438-1446. DOI: 10.1016/j.matdes.2007.09.002

[15] Chethan KN, Bhat SN, Zuber M, Shenoy SB. Patient-specific static structural analysis of femur bone of different lengths. The Open Biomedical Engineering Journal. 2018;**12**:108-114. DOI: 10.2174/1874120701812010108

[16] Evans FG. Mechanical Properties of bone. Artificial Limbs. 1969;**13**(1):37-48

[17] Bessho M, Ohnishi I, Matsumoto T, Ohashi S, Matsuyama J, Tobita K, et al. Prediction of proximal femur strength using a CT-based nonlinear finite element method: Differences in predicted fracture load and site with changing load and boundary conditions. Bone. 2009;**45**(2):226-231. DOI: 10.1016/j.bone.2009.04.241

Chapter 7

A Review of Significant Advances in Areca Fiber Composites

Narayanan Gokarneshan, Venkatesan Sathya, Jayagopal Lavanya, Shaistha Shabnum, Habeebunisa and Sona M. Anton

Abstract

This chapter provides a comprehensive review of the recent developments in the design of areca fiber composites. The physical, mechanical, and thermal properties of areca fiber and its composites are explained here. The species of Areca fiber represents the Arecaceae/Palmae family (like the coconut/palm trees), with regard to its physical and mechanical properties. Researchers identified that areca fiber holds prospective applications as an alternative to reinforced polymer composites in the automotive, aerospace, and construction industries. Surveys on bio-softening, adhesion, the effects of fiber length, chemical treatments of long areca fibers, the influence of mercerization on the tensile strength of long and short areca fibers, and areca husk have been done. Several researchers have utilized various natural fibers in developing bio-composites. Furthermore, the reinforced composite of natural fiber is a prospective research area, considering its mechanical properties, tensile strength, lightweight, nominal pricing, biodegradable/eco-friendly nature, and ease of procuring raw materials compared to synthetic fiber-reinforced composites. However, little research has been done on areca leaf fibers as a feasible fiber. This chapter provides information on the development and investigation of the mechanical behaviour of a natural fiber-reinforced epoxy composite of areca fiber with various configurations of areca fiber orientation.

Keywords: areca fibers, natural fibers, mechanical properties, hybrid composites, chemical treatment, thermal properties

1. Introduction

Synthetic fibers have been found to show excellent properties that describe their wide areas of application in different industries. But such fibers can pose environmental issues, with regard to landfills, owing to their non-biodegradability [1]. Natural fibers have been intended to be used in composites owing to environmental considerations in the form of individual or hybrid reinforcement fibers. Prior to being utilized in composites, natural fibers have been obtained from nature in the form of animals, plants, and minerals, using different techniques, like chemical and thermal modifications.

Areca fiber is chiefly obtained from the fruit areca, frond, and stalk leaf. The world today is facing the problem of growing new and propelled innovations and methods to eliminate or utilize solid wastes, particularly with polymers that are non-reversible in nature. The methods adopted in splitting up the wastes do not seem economical and tend to generate chemicals that prove harmful. Taking into account these factors, reinforcing polymers using natural fibers seems the only option that could result in solving the issue. Regular strands are easily available and reusable, have less thickness, and are ecofriendly. They possess high tensile properties and can be used to substitute the customary strands. The strong demerit of using characteristic strands for strengthening plastics is the contrariness, causing weak bondage between normal filaments and lattice gums and thus leading to low pliable characteristics. A number of theories and surface modification methods have been evolved to improve fiber-network interfacial holding and enhance malleable characteristics of the composites. Further, it is proved that the strength and stiffness of the natural fiber polymer composites are mainly influenced by the loading of fiber. Up to a particular extent, there is a rise in mechanical properties with increasing fiber weight ratio. In order to evaluate the tensile properties of natural fiber reinforced composites, mathematical models/finite element models are being adopted as a necessity.

Natural fiber comprises cellulose, lignin, pectin, and so on. Owing to the presence of such constituents, natural fiber possesses unique features and special properties and gives high moisture percentage, which would in turn influence the fiber–matrix bonding. In order to find a solution for this problem, certain techniques of chemical treatment have been evolved and investigated so as to satisfy the properties of other man-made fibers [2–6]. When considering end uses like electrical insulation, the areca/betel nut fiber reinforced composites exhibit higher merits with regard to the latest development of composite materials [7].

The requirements of high strength to weight ratio in components prompted the development of composites, which necessitated high performance and efficiency, and in turn led to advances in different polymer matrix composites having different fiber reinforcements like carbon fiber, glass fibers, aramid, natural fibers, hybrid, and so on. Natural fiber composites assume a crucial role taking into account the factor of environment-friendly materials and the necessity to manufacture different sustainable engineering and industry-oriented components.

Owing to their good mechanical properties and biodegradability, natural fibers have a crucial function as a reinforcement agent and are readily available in many parts of the world. A number of natural fibers such as jute, kenaf, sisal, hemp, bamboo, areca, pineapple, banana, and coir are being considered important for several research studies due to their availability and cost effectiveness for the design of a cost-effective reinforcing material [8–11]. A number of properties arise from the use of various natural fibers as a reinforcement agent in composite materials and can effectively be utilized for different end uses.

2. Evaluation of the physical, thermal, and mechanical properties

The different parts of plants like bast, leaf, seed, stalk, fruit, grass, and wood yield cellulosic or lignocellulosic fibers. Fruits of plants yield fibers that are normally short, light, and hairy; bast (found in the stem or trunk) yields long fibers that offer strength to the plant or tree. Sturdy and rough fibers are obtained from leaves and are normally utilized in the transportation and automotive sectors. Fiber length is considered important for use of the fiber, particularly in the traditional fiber industries [12]. High-quality fabrics can be designed from yarns spun by long fibers (clothing, laces, domestic textiles, tents, sailcloth), whereas fibers like cotton, flax, hemp, ramie, and sisal can be used for production of coarser fabrics such as bagging, floor coverings, and carpets. Fibers such as jute, sisal, cotton, and hemp can be considered for cordage fiber, tying twine, rope, and binder twine [13, 14]. Also, sisal and coir fibers can be utilized in brushes and for weaving to produce hats, mats, baskets, and rugs [15]. Such fibers have also been utilized as fillers in upholstery, for seams in vessels, barrels, and piping and as reinforcement for plastic and wallboard.

Moreover, natural fibers can be used with wood pulp in manufacturing paper [16]. Investigations on natural fibers, particularly kenaf, jute, and bamboo, have increased over the past few years [17–23]. For instance, an investigation relating to the ballistic impact resistance of kenaf reinforced polyvinyl butyral composites; a study of flexural strength and ductility of kenaf reinforced concrete composites; work relating to the influence of kenaf hybridization with oil palm fiber reinforced an epoxy matrix on the tensile, flexural, and impact properties of the obtained composites; and research with regard to processing and manufacturing of kenaf reinforced that epoxy composites are worthy of consideration [24–26].

Studies have been carried out on fiber hybridization relating to kenaf and fiberglass to find out its influence on the tensile and impact properties of the materials so produced [27]. Besides, a number of workers also reported on the use of natural fibers in the design of industrial safety helmets [28].

Previous investigations on natural fibers have shown scanty research carried out on areca and other species from the Palmae family having identical properties. Despite the abundance of areca palm in South East Asia and the Pacific region, its fiber has not still attracted much attention and is presently being less used than other palm tree fibers [19].

It has been found that less substantial research has been conducted with regard to the optimization of surface treatment, production technique, and application of areca fiber as a reinforcing material in composites. At present, very little literature is available on areca fiber used as reinforcement in composites, which implies that in spite of its innumerable merits, the fiber is at present used less. The fiber enjoys merits like recyclability; renewability; sustainability; economy; wide availability; high-potential perennial crop; inherited qualities; superior properties; mechanical properties that compare well with those of other fibers like kenaf, jute, and coir; and also complete biodegradability [29]. Statistics shows that the annual world production of areca nuts is 1,073,000, and approximately 2.5 g of areca husk could be extracted from every areca betel nut [9]. The annual statistics on world natural fiber production shows the least production of areca husk fibers in comparison with other natural fibers, like jute, coir, and kenaf, which could be ascribed to the consumption of betel nuts in the tropical Pacific and Eastern Africa and Asia.

Areca catechu is known by various names like areca palm, areca nut palm, and betel palm. It is also called Pinang in Malaysia. *A. catechu* is found largely in the tropical Pacific, Eastern Africa, and Asia, particularly in Malaysia, Philippines, India, and Sri Lanka. As per the statistical data provided by the Food and Agriculture Organization of the United Nations, India, Myanmar, Bangladesh, China, and Indonesia are considered the major producers of betel nut [30]. Sri Lanka and India are the two countries where *A. catechu* trees are well grown, and the people of these countries use betel nut as a complement to betel leaves smeared with limestone paste [31]. But betel nut fibers are used as housing insulation material in a traditional way in certain countries [32]. Areca fruit finds a number of medical applications that include dental implants, drugs for wounds, healing of sores, diphtheria, heavy menstrual blood flow, diarrhea, and ulcers [33]. On the other hand, biodegradable disposable plates are made from areca leaf sheaths, which fall naturally from the trees, or green waste. As the use of plastic is banned in India, areca plates are widely used. Besides India, other countries including China, Vietnam, Ukraine, Sri Lanka, Malaysia, and the United Arab Emirates manufacture these plates.

The areca tree can reach a height of 10 to 20 m, with an erect stem that is single and thin, having a diameter ranging between 10 to 15 cm with impressions of annulated scars of fallen leaf sheaths or fronds.

The leaves span a length between 150 and 200 cm; having many pinnate-shaped leaves, the upper part normally shows 8 to 12 fronds. Fully grown areca trees measure up to 15 m. But the conditions of soil mainly influence the growth of such trees [34, 35]. *A. catechu* is a monocotyledonous plant that belongs to the species of the *Areca* and plant family of *Arecaceae* or *Palmae* [36]. It relates to the species of oil palm, date palm, coconut palm, and others. On the whole, the plants from the Palmae family can be considered tropical trees, shrubs, and vines, normally with a tall columnar trunk, bearing a crown of huge leaves. Many investigations have been carried out on the use of plant fibers extracted from the plant family. They point to the prospect of *A. catechu* fibers to be used as an option as reinforcement in natural fiber-based composites [37–40].

2.1 Thermal properties of areca fibers

In the design of natural fiber composites, thermal stability is considered crucial. It can decide the selection of compatible processing techniques for fibers and composites. Hence, thermal properties act as a guideline during the entire design process and prevent the temperature from rising above the degradation temperature of the fiber, since it could decrease the performance of the fibers and the composites.

The fiber is found to be thermally stable up to 230°C, as evinced by lack of weight loss after the minor loss caused by moisture evaporation. Beyond this point, there is occurrence of polymerization and degradation processes of hemicelluloses and cellulose up to 330°C. Analysis of the DTG curve shows small peaks at 273.4 and 325.8°C and reveals the pyrolysis, decomposition, and degradation of hemicelluloses and cellulose. It is found that the kinetic activation energy for areca fibers falls in the range set for natural materials.

The value is indicative of areca fibers possessing excellent thermal stability, which permits it to undergo the polymerization process in the production of composites. At a temperature of about 325°C, the burning of fiber has been evinced, which is a reasonably high temperature for polymer processing to manufacture composites.

2.2 Mechanical properties of areca fiber

Single fiber tensile testing has been used to evaluate the mechanical properties and provide some basic information necessary for the design of the potential use of plant fibers. Areca fibers have been compared with coir and palm leaf fibers with regard to the mechanical properties, particularly tensile strength. This could be attributed to its high crystallinity index and spiral angle. Considering application in reinforcement, the greater strain and low modulus of areca husk fiber offer superior toughness. The results indicate that areca fiber can substitute reinforced polymer composites, similar to other representatives from the family of *Palmae/Arecaceae*.

On the other hand, chemical modification also determines the mechanical properties of the fiber. The untreated and alkali-treated fibers in selected concentration and weight have been characterized, and the changes undergone by the removal or minimization of non-cellulose components, like hemicellulose, lignin, and wax pectin, and other impurities from the fiber surface have been described [41–43]. The modification results in surface roughness and fibrillation due to the exclusion of cementing materials that lead to improved mechanical properties of the fiber reinforced composite [42]. 5% alkali-treated fiber has been found to exhibit the greatest tensile strength and modulus based on tensile characterization of untreated and treated areca leaf stalk fibers. This could be attributed to the disruption of hydrogen bonds in the fiber network. But a reduction in tensile strength has been observed with a rise in alkali concentration above the optimum. The tensile strength of individual fibers has been enabled by the increase in the pores and pits on the surface of fibers. Also, benzoylation treatment is found to yield better tensile properties in comparison with the alkali-treated and untreated short areca sheath fibers. The FTIR studies reveal that the absorption of alkali and benzyolation treatments have decreased the (–OH) groups compared to in the case of the untreated fiber, due to the removal of hemicelluloses. Further, the presence of phenyl nucleus has been noticed, while the C – H deformation from lignin confirmed its removal, and the aromatic ring associated with the C – O bond demonstrated the removal of hemicellulose and pectin.

3. Fabrication of natural fiber reinforced composites

3.1 Composites reinforcement of the fibers in the polymeric matrix to improve strength, stiffness, and fatigue resistance

A number of conventional methods have been adopted for the production of natural fiber reinforced composites. Among these, the pultrusion technique is apt for pulling continuous fiber rovings through a resin bath by using a puller. But filament winding is suitable for the product in the circular form of continuous fiber roving drawn through a resin bath by a puller. In the case of hand lay-up or wet lay-up, each layer of fiber is wetted by the resin with a roller to consolidate. It is an open molding process. A closed vessel or pressure gradient is used in an RTM, for streaming and then soaking with reaction resin the long or woven fibers. In the case of natural fibers, compression molding can be used, with fibers that are continuous or discontinuous, accompanied by temperature, pressure, and high volume. For preparation of thermos-setting resin, sheet molding compound (SMC) and bulk molding compound (BMC) have been commonly adopted, whereas glass mat thermoplastic (GMT) has been adopted in the case of thermoplastic polymer. Injection molding can be adopted in the preparation of polymer composites using short natural fibers in particle or powder form. The molten composites are injected into the mold, which is ejected after cooling. The fabrication selection of natural fiber reinforced polymeric composites is chosen on the basis of a few factors considered.

Very little research has been reported on the areca fiber reinforced polymer composites. Such composites are made by areca fiber reinforced polyester through utilization of hand lay-up, cured for 1 day at room temperature [37]. Compression at a temperature of 115°C for 30 min is used for the production of areca fiber mats with PLA laminate composites [44]. But hand lay-up has been used for 300–325 μm fabricated areca fiber reinforced vinyl ester resin and oven-cured at 80°C for 2 h [45]. The areca fiber reinforced polypropylene composites have been prepared by extrusion and hot molding press [46].

3.2 Mechanical properties of areca reinforced polymeric composites

The mechanical properties of areca reinforced polymer composites with different fiber loading and matrixes have been determined. The influence of adding untreated and treated betel husk fibers, in various percentages, to polypropylene composites has been studied [46]. The findings reveal that tensile, bending, and impact properties have been excellent with 30 wt% fiber loading. Hence, the formulation has been selected for further study, comprising washing with detergent and alkali for surface treatment. In comparison with untreated and detergent-washed fibers, alkali-treated fiber showed excellent properties. However, a 40 wt% areca husk fiber reinforced unsaturated polyester composite has been observed to give enhanced mechanical properties. On the other hand, a rise in the fiber loading resulted in fiber pull-out and debonding and decreased the load-bearing capacity of the composite. It has been found that chemically treated areca reinforced composites showed better mechanical properties, like tensile, flexural, and impact strength [47]. It can be described on the basis that the chemical treatment removes impurities, like pectin, fat, and lignin, from the natural fiber and subsequently improves the fiber-matrix bonding and enhancement in the mechanical properties of the composite. Also, regression studies carried out to assess the mechanical properties of untreated and treated short areca sheath reinforced polyvinyl alcohol (PVA) thermoplastic matrix have shown that 27 wt% fiber loading reveals optimum mechanical properties. It is found that benzyl chloride treatment is the perfect alternative to the various prospects of surface treatment in the case of areca sheath fiber [48]. Also, the influences of various chemical modifications in removing impurities (pectin, fat, lignin) from areca husk fiber at different compositions have been further used in polypropylene for enhancement of its mechanical properties. The results strengthen the fact that caustic, potassium permanganate, benzoyl chloride, and acrylic acid treatments improve the mechanical properties of the fibers. Depending on the studies of flexural, tensile, and impact strength properties, it has been observed that the 60 wt% fiber formulation gave the greatest values in flexural and tensile testing, whereas the 50 wt% fiber formulation showed the best effects for impact strength. The treatment with acrylic acid has been observed to be the most suitable chemical modification and has been the most effective in improvement of the tensile, flexural, and impact properties of the areca husk fiber reinforced polypropylene composites. Also, the physical, mechanical, and thermal properties of betel nut husk fibers extracted from raw fruit and used to reinforce an epoxy matrix have been studied.

Fiber percentages of different concentrations have been incorporated into the composites, and it has been found that 5% fiber loading yielded the highest tensile strength and hardness. It has been found that there is a degradation of thermal and mechanical properties at excessive fiber loading of 8% fiber content. It is caused by a lack of interface bonding between fiber and matrix that created a disruption in load transfer. It is observed that the incorporation of 10 wt % betel nut husk fiber in a vinyl ester matrix considerably decreased the flexural strength and also added to a high increment in the tensile and impact properties, mainly for the fiber extracted from raw and ripe areca fruit. In the case of matured fruit, the tensile and impact properties of the extracted fiber have been found to be lower in comparison with the matrix, due to the presence of high lignin content, which influenced brittle fracture.

However, composites have been designed on untreated short areca sheath fibers with varying proportions of fiber weights of fiber reinforced polypropylene composites, using hot and cold compression molding, and it has been observed that the

10 wt% fiber formulation gave excellent tensile strength and the best weathering resistance for the formulations investigated. Likewise, random fiber orientation with the hand lay-up method has been used to prepare the 30 vol% untreated areca sheath fiber reinforced epoxy resin composites, so as to assess the flexural strength by 3- and 4-point bending tests. The findings have been validated by the numerical simulation method.

The formulation is useful for end uses that need less load bearing structural and non-structural capacity [49]. The compatibility of areca fiber and the matrix has been studied, and it has been reported that the interfacial bonding and adhesion bonding between the fiber and the matrix can result in better mechanical properties and good stress transfer.

Just as other natural fibers, areca fiber also exhibits certain demerits, particularly with reference to compounding difficulties owing to the inherently polar and hydrophilic nature of the fiber that results in non-uniform distribution, thus weakening the properties of the composite. During the production of composites, the fiber degradation that occurs at the processing temperature of the matrix imposes another practical constraint. Fiber wettability is another aspect to be considered, which affects its compatibility with the matrix. The surface tension and matrix viscosity are determined by wettability.

Hence, it is imperative that the surface tension of the reinforcing fiber exceeds that of the matrix so as to maintain the interfacial strength. The other aspects to be taken into account are low microbial resistance and susceptibility to rotting so as to attain the successful preparation of a long-lasting composite. The bonding of the fiber matrix and surface wetting gets affected due to the natural waxy substance on the surface of the fiber [50].

3.3 Areca-based hybrid composites

The hybridization of different fibers, of natural or artificial origin, that are reinforced in a same polymer matrix results in a hybrid composite. In the case of reinforced polymer composites, the hybridization of areca fibers with other fibers has been considered in this case. Compression molding has been used for the production of hybridization of areca husk-coir fiber reinforced unsaturated polyester and areca husk-sisal fiber reinforced epoxy composites [51, 52]. On the other hand, the hot press method has been adopted to produce areca husk-glass fiber reinforced polyethylene composites.

3.4 Mechanical properties of areca-based hybrid composites

Explanation has been provided regarding the hybridization of areca fibers with other fibers for the manufacture of reinforced polymer composites. Compression molding technique has been adopted in the hybridization of areca husk-coir fiber reinforced unsaturated polyester and areca husk-sisal fiber reinforced epoxy composites [53]. On the other hand, the hot press method has been adopted in the production of areca husk-glass fiber reinforced polyethylene composites. Taking into account the drastic rise in the number of plastic wastes, which create an adverse effect on the ecosystem, environmental considerations have prompted a push in bio-composite studies. Studies have been conducted on the tensile, flexural, impact, and hardness of composites made by 20 wt% natural fiber loading comprising 10 wt% caustic-treated areca fiber, coir fiber, or a mixture of both dispersed in a polyester matrix.

The hybridization of natural fibers to reinforce polymer composites results in improved mechanical performance, in comparison with individual fiber-based polymer composites [54]. But it has been observed that, in certain instances, individual fiber-based polymer composites have shown superior mechanical properties as compared with hybrid composites because of the micromechanical behavior of the individual components of the reinforcement [55]. The areca husk reinforced polyester composites have been observed to be superior in comparison with hybrid areca/coir fiber reinforced polyester composites. Another research reported on 5% caustic treated areca fiber hybridized with glass fiber, in selected proportions, having 20% fiber loading in a thermoplastic matrix of polyethylene, and it has been investigated so as to study the effect of surface modification on the physico-mechanical properties of the composite. Depending on the comparison with the properties of polyethylene considered as a standard test sample, it has been found that the tensile and flexural properties have considerably improved for the formulations having selected proportions. The areca fiber coated with caustic showed a considerable enhancement in mechanical properties. With the rise in proportion of the glass fiber three times, it showed an improvement in tensile strength, Young's modulus, flexural strength, and hardness but not in impact energy.

A work has reported on a study of the tensile, bending, and impact testing results for hybrid areca and jute reinforced epoxy composites, made with the hand lay-up method and consisting of a tri-layer, having areca fibers as surface layers and jute as a central layer. The tensile and bending strengths obtained at different levels, with epoxy LY 556 as the control specimen, have shown that the hybridization of areca and jute, with the epoxy matrix, considerably decreased the tensile and bending strengths of the composites.

However, work has been carried out on the hybridization of areca husk and sisal fiber, with a similar kind of epoxy, and it is observed that there is a slight decrease in tensile strength. However, there has been drastic rise in the flexural strength. It has been observed that there is a rise in the tensile and flexural strengths due to an increase in the content of betel nut fiber rather than that of sisal fiber. The effect of chemical modification with NaOH for composites comprising 20% areca frond, 20% sisal, and 60% of the same epoxy has been studied, and the findings show that the caustic-treated composites have revealed improved mechanical properties, like tensile and flexural strength, absorbed energy, and hardness, in comparison with the untreated specimen [56]. It has been observed that the entire spectrum of the composite properties is considerably influenced by the fiber content [57].

4. Evaluation of mechanical properties

Despite the modulus being dependent on the fiber properties, the tensile strength is sensitive to the grid properties. It is necessary to have a solid interface, low anxiety fixation, and fiber introduction so as to improve the strength. On the other hand, the elastic modulus is determined by fiber concentration [5–7, 58].

For certain tribological uses, treated betel nut fiber reinforced polymer composites have been regarded better in comparison with chopped strand mat glass fiber-reinforced polyester; betel nut polyester composite possesses mechanical properties identical to those of the glass-polyester composite. Thus, betel nut fibers possess a great prospect to replace glass fibers and for small load applications [59–61].

With regard to the high tensile strength applications, chemically treated areca fiber reinforced natural rubber composite, and also uses related to high dimensional stability, low-density property of raw betel nut husk fiber is used for lightweight applications [60–62]. Natural fibers have economy and biodegradability, can be reutilized, and are eco-friendly materials. Natural fibers can be a better choice over glass and carbon fibers owing to their eco-friendliness and biodegradable nature. Betel nut and *Sansevieria cylindrical* in PP (polypropylene) composites have found applications where strength and cost considerations are important [63].

Likewise, areca fiber and maize powder reinforced PF composites have been used in packing industries, low-cost housing, and domestic uses [64]. Areca sheath fiber is used in structural and non-structural areas like suitcases, post boxes, grain storage, automobile interiors, partition boards, and indoor uses [65].

4.1 Uses

From the aforesaid explanations, uses of naturally available, eco-friendly, renewable and reproducible, nontoxic, economical, and easily available reinforcing material (areca/betel nut fibers) composites, it can be summarized that areca fiber offers a good alternative for wood in indoor uses, and the following points have been listed that are present in the literature. Due to a rise in the volume fraction of fiber in the composite, there is a rise in dielectric strength of betel nut (Bn) composites. It is a rather uncommon phenomenon that has not been noticed in a number of natural fiber composites. Hence, based on the availability, cheaper and good dielectric strength of Bn fiber composite can certainly be considered for electrical insulation applications [66]. Further, hybrid composites with Bn and *S. cylindrica* in EP can be used in diverse applications as structural materials.

- Depending on the availability, cost-effectiveness, and good strength of areca fiber composites, they are utilized in the design of lightweight materials that are used in automobile body building, office furniture packaging industry, partition panels, and others compared to wood-based plywood or particle boards [67–69].

- Due to better outcomes with regard to wear of treated betel nut fiber reinforced polymer composites (T-BFRP) (98%) in comparison with chopped strand mat glass fiber reinforced polyester (CSM-GFRP) under dry and wet states, T-BFRP composite holds promise in certain areas of tribological uses.

- With regard to mechanical properties, the betel nut polyester composite is identical to the glass-polyester composite, and thus, betel nut fibers offer a very good substitute for glass fibers in mechanical end uses. Also, betel nut fibers offer superior support to the polyester matrix than other types of natural fibers.

- In the case of end uses related to a small load bearing, investigations have been conducted on caustic-treated areca fiber composites, which have been found to have enhanced mechanical properties to a certain extent in the areca fibers.

- In the case of end uses requiring high tensile strength, chemically treated areca fiber reinforced natural rubber composites have been considered.

- For end uses requiring high dimensional stability, low moisture and water uptake properties of dried BNH fibers are found to be advantageous for various applications, and low-density property of raw BNH fiber is used in applications requiring light weight.

- Certain investigations make it evident that areca fruit husk fibers are useful as a potential reinforcement in polymer composites due to their moderate tensile strength properties, better strength, and bonding properties with rough surface morphology in end uses requiring light weight.
- Hybrid composites with betel nut and *S. cylindrical* in PP (polypropylene) can be used in various areas like structural materials that are dictated by strength and cost aspects.
- In areas such as packing industries, low-cost housing, and domestic purposes, areca fiber and maize powder reinforced PF composite materials can be used as a communicative material for plywood.
- In the case of structural and non-structural uses like suitcases, post boxes, grain storage, automobile interiors, partition boards, and indoor uses, untreated chopped natural areca sheath fiber reinforced polymer matrix bio-composites are well adopted.
- In the case of end uses requiring high flexural strength, chemically treated areca fiber reinforced epoxy composites for applications where high impact strength is required, 60% fiber loading was considered.

5. Characterization

Natural fiber reinforced composites are found to offer resistance to electricity, possess good thermal insulating property, and are also resistant to corrosion [70]. It is found that as the volume fraction of the fiber increases, the tensile strength of natural polymer composites increases, as pointed by past research works on volume fraction [71]. It has been shown in the study of the tensile characteristic of the untreated areca sheath composite that the longer the fiber length, more is the strength of the composite. In comparison with untreated areca nut fibers, the fibers treated with caustic exhibit greater mechanical strength [72, 73]. The curing time of the composite also plays a vital role in determining the full potential strength of the composite. Studies have revealed that the strength of the composite increases with the curing time [7]. Investigations have been carried out with regard to other mechanical properties like impact strength, hardness, and flexural strength of the areca fiber composite, considering parameters like the influence of volume fraction, post curing time, and alkali treatment for effective bonding [14, 74]. The flexural strength increases with increase in fiber loading percentage. Fibers treated with caustic show a significant increase in flexural strength for references [8, 9, 68, 75]. Studies show that there is a rise in the impact strength due to the post-curing time, and it has been observed that with the rise in the curing time, the alkali-treated composites turned brittle [10, 57, 76, 77]. The finding aims on the tensile property of the non-chemically treated areca fiber sheaths that are immersed in water, hot pressed, and compressed with a thickness of 2.5 to 3 mm and are widely found in India as areca fiber plates. The study also aims to use this as a reinforcing agent in epoxy polymer composite to study its tensile properties and effects of the different combinations of the orientation angles of the fiber, which influences the tensile property of the composite.

6. Conclusion

Studies have been carried out with regard to the physical, mechanical, and thermal properties of areca fiber and its composites. The available literature shows that areca fiber holds the prospect for use as an optional reinforcement in polymer composites. On the other hand, it has been found that certain other species from the family of Arecaceae/Palmae have not been studied since natural fiber reinforcement in polymer composites either, hence showing further options of natural alternative reinforcements to be studied in terms of their potential for utilization in fiber-based composites for the automotive, aerospace, and construction sectors. Areca fibers can be used as an alternative natural resource and as a promising reinforcement of polymer matrices to produce lightweight composite structures. Different natural fibers have been used by many researchers for the development of bio-composites, but areca leaf fibers as a feasible fiber have seldom been researched or spoken about.

Author details

Narayanan Gokarneshan[1*], Venkatesan Sathya[2], Jayagopal Lavanya[1], Shaistha Shabnum[3], Habeebunisa[4] and Sona M. Anton[5]

1 Department of Fashion Design, Hindustan Institute of Technology and Science, Chennai, India

2 Department of Fashion Design, SRM Institute of Technology and Science, Ramapuram, Chennai, India

3 Interior Design and Decor, Department of Home Science, JBAS College for Women, Chennai, India

4 Department of Knitwear Design, National Institute of Fashion Technology, Chennai, India

5 MET College, Nagercoil, India

*Address all correspondence to: advaitcbe@rediffmail.com

References

[1] Thakur S, Chaudhary J, Sharma B, Verma A, Tamulevicius S, Thakur VK. Sustainability of bioplastics: Opportunities and challenges. Current Opinion Green Sustainability Chemistry. 2018;**13**:68-75

[2] MohanKumar GC. WCE. Proceedings of the World Congress on Engineering, London, UK. July 2008 vol. II, 2-4

[3] Rajan A, et al. Bioactive Polymer Engineering Section, RRL (CSIR), Trivandrum 695019, India. 2005

[4] Hazarika SB, et al. Natural fibre reinforced polyester based biocomposite: Agro waste utilization. International Journal of Scientific and Industrial Research. 2015;**74**:589-594

[5] Srinivasa CV, et al. Water Absorption of Areca Fiber Reinforced polymer composites. Advances in Polymer Technology. 2012;**31**:319-330

[6] Padmaraj NH, et al. Proceedings of the IV National Conference on Trends in Mechanical Engineering, TIME'10 (30th December 2010). 2010. pp. 30-33

[7] Ramachandra Reddy G et al. Fabrication and performance of hybrid Betel nut (Areca catechu) short fiber/Sansevieria cylindrica (Agavaceae) epoxy composites. International Journal of Materials and Biomaterials Applications. 2011;**1**(1):6-13

[8] Taj S, Munawar MA, Khan S. Natural fiber-reinforced polymer composites. Journal Proceedings of the Pakistan Academy of Sciences. **44**(2):129-144. ISSN 0377-2969

[9] Desai RH, Krishnamurthy L, Shridhar TN. Effectiveness of areca (betel) fiber as a reinforcing material in eco-friendly composites: A Review. Indian Journal of Advances in Chemical Science. 2016;**S1**:27-33

[10] Devi LU, Bhagawan SS, Thomas S. Polyester composites of short pineapple fiber and glass fiber: Tensile and impact properties. Polymer Composites. 2012;**33**:1064-1070

[11] Jeyasekaran AS, Kumar KP, Rajarajan S. Numerical and experimental analysis on tensile properties of banana and glass fibers reinforced epoxy composites. Sādhanā. **41**(11):1357-1367

[12] Saba N, Md Tahir P, Jawaid M. A review on potentiality of nano filler/natural fiber filled polymer hybrid composites. Polymers. 2014;**6**:2247-2273

[13] Saba N, Paridah MT, Jawaid M. Mechanical properties of kenaf fiber reinforced polymer composite: A review. Construction and Building Materials. 2015;**76**:87-96

[14] Wigglesworth E. Thread, twine, cordage, and rope. Journal of the Textile Institute Proceedings. 1925;**16**:139-142

[15] Hannan WI. The Textile Fibers of Commerce, C. Griffin, London; 1902. pp. 167-168

[16] Thompson G, Swain J, Kay M, Forster C. The treatment of pulp and paper mill effluent: A review. Bioresource Technology. 2001;**77**:275-286

[17] Hanan F, Jawaid M, Md TP. Mechanical performance of oil palm/kenaf fiber-reinforced epoxy-based bilayer hybrid composites. Journal of Natural Fibers. 2018;**17**:1-13

[18] Bajpai PK, Ram K, Gahlot LK, Jha VK. Fabrication of glass/jute/epoxy composite based industrial safety

helmet. Material Today: Proceedings. 2018;**5**:8699-8706

[19] Nor AFM, Sultan MTH, Hamdan A, Azmi AMR, Jayakrisna K. Hybrid composites based on kenaf, jute, fiberglass woven fabrics: Tensile and impact properties. Material Today: Proceedings. 2018;**5**:11198-11207

[20] Alshammari FZ, Saleh KH, Yousif BF, Alajmi A, Shalwan A, Alotaibi JG. The influence of fiber orientation on tribological performance of jute fiber reinforced epoxy composites considering different mat orientations. Tribology in Industry. 2018;**40**:335-348

[21] Kumar R, Bhargav C, Bhowmik S. Bamboo fiber reinforced thermoset and thermoplastic polymer composites: A short review. AIP Conference Proceedings. 2018;**1998**:020018

[22] Md Shah AU, Sultan MTH, Jawaid M, Cardona F, Talib ARA. A review on the tensile properties of bamboo fiber reinforced polymer composites. BioResources. 2016;**11**:10654-10676

[23] Osorio L, Trujillo E, Van Vuure AW, Verpoest I. Morphological aspects and mechanical properties of single bamboo fibers and flexural characterization of bamboo/epoxy composites. Journal Reinforcement Plastics and Composites. 2011;**30**:396-408

[24] Salman SD, Leman ZB. Physical, mechanical and ballistic properties of kenaf fiber reinforced poly vinyl butyral and its hybrid composites. In: Editors, Sapuan SM, Ismail H, Zainudin ES, editors. Natural Fiber Reinforced Vinyl Ester and Vinyl Polymer Composites. USA, Philadelphia: Woodhead Publishing; 2018. pp. 249-263

[25] Saba N, et al. Mechanical properties of kenaf fibre reinforced polymer composite: A review. In: Michael CF, Masayusa O, Muhammed Basheer PA, Kent AH, editors. Construction and Building Materials. Oxford-UK: Elsevier Ltd; 2015:87-96

[26] Syed Mohsin SM, Baarimah AO, Jokhio GA. Effect of kenaf fiber in reinforced concrete slab. Material Science and Engineering Conference Series. 2018;**342**:012104

[27] Saba N, Paridah MT, Jawaid M, Abdan K, Ibrahim NA. Manufacturing and processing of kenaf fiber-reinforced epoxy composites via different methods. In: Salit M, Jawaid M, Yusoff N, Hoque M, editors. Manufacturing of Natural Fiber Reinforced Polymer Composites. Berlin, Germany: Springer, Cham; 2015. pp. 101-124. DOI: 10.1007/978-3-319-07944-8

[28] Jitendra KP, et al. Recent advances in the application of natural fiber based composites. Macromolecular Materials and Engineering. Weinheim, Germany: WILEY-VCH Verlag GmbH & Co. KGaA; 2010;**295**:975-989

[29] Ihwah A, Deoranto P, Wijana S, Dewi IA. Comparative study between Federer and Gomez method for number of replication in complete randomized design using simulation: Study of Areca Palm (Areca catechu) as organic waste for producing handicraft paper. Earth Environmental Science. 2018;**131**:012049

[30] Purushotham B, Narayanaswamy P, Simon L, Shyamalamma S, Mahabaleshwar H, Jayapalogwdu B. Genetic relationship between cultivars of areca nut (Areca catechu L.) determined by RAPD. The Asian and Australiasian Journal of Plant Science and Biotechnology. 2008;**2**:31-35

[31] FAOSTAT F. Food and Agriculture Organization statistical database Retrieved Feb. 2014, Available online: http://faostat.fao.org/

[32] FAO Statistical Pocketbook, Food and agriculture organization of the United Nations. 2017:231. Available from: https://www. fao.org/3/i4691e/i4691e. pdf. ISBN: 978-92-5-108802-9

[33] Bodeker G, Salleh H, Shekar S. Health & Beauty From the Rainforest: Malaysian Traditions of Ramuan. Kuala Lumpur, Malaysia: Editions Didier Millet Pty Ltd; 2009

[34] Haque MM, Hasan M. Influence of fiber surface treatment on physico-mechanical properties of betel nut and glass fiber reinforced hybrid polyethylene composites. Advanced Material Process Technology. 2018;**4**:511-525

[35] Patil PR, Rakesh SU, Dhabale PN, Burade KB. Pharmacological activities of Areca catechu Linn – A review. Journal of Pharmacy Research. 2009;**2**:683-687

[36] Binoj JS, Raj RE, Sreenivasan VS, Thusnavis GR. Morphological, physical, mechanical, chemical and thermal characterization of sustainable indian areca fruit husk fibers (Areca Catechu L.) as potential alternate for hazardous synthetic fibers. Journal Bionic Engineering. 2016;**13**:156-165

[37] Chiduruppa M, Ahmad F, Pandian P. A detailed review on dypsis lutescens (Arecaceae). World Journal of Pharmacy and Pharmaceutical Sciences. 2018;**7**:1644-1650

[38] Ashok RB, Srinivasa CV, Basavaraju B. A review on the mechanical properties of areca fiber reinforced composites. Science Technology Materials. 2018;**30**:120-130

[39] Yusriah L, Sapuan SM. Properties of betel nut husk forced vinyl ester composites. Natural Fiber Reinforced Vinyl Ester and Vinyl Polymer Composites. Elsevier Ltd. 2018:129-155. DOI: 10.1016/B978-0-08-102160-6.00006-8

[40] Nayak S, Mohanty JR. Study of mechanical, thermal, and rheological properties of Areca fiber-reinforced polyvinyl alcohol composite. Journal of Natural Fibers. 2018;**16**:1-14

[41] Shanmugasundaram N, Rajendran I, Ramkumar T. Characterization of untreated and alkali treated new cellulosic fiber from an Areca palm leaf stalk as potential reinforcement in polymer composites. Carbohydrate Polymer. 2018;**195**:566-575

[42] Oushabi A, Sair S, Oudrhiri Hassani F, Abboud Y, Tanane O, El Bouari A. The effect of alkali treatment on mechanical, morphological and thermal properties of date palm fibers (DPFs): Study of the interface of DPF–Polyurethane composite. South African Journal of Chemical Engineering. 2017;**23**:116-123

[43] Mwaikambo LY, Ansell MP. Chemical modification of hemp, sisal, jute, and kapok fibers by alkalization. Journal of Applied Polymer Science. 2002;**84**:2222-2234

[44] Kousaalya AB, Biddappa BI, Krumm K, Pradeep SA, Pilla S. Poly(lactic acid)/areca fiber laminate composites processed via film stacking technique. Journal of Applied Polymer Science. 2018;**135**:45795-45807

[45] Lazim Y, Salit SM, Mariatti M, Zainudin ES, Mustapha M, Jawaid M. Effect of alkali treatment on the physical, mechanical, and morphological properties of waste betel nut (Areca catechu) husk fiber. BioResources. 2014;**9**:7721-7736

[46] Chakrabarty J, Hassan MM, Khan MA. Effect of surface treatment on betel nut (areca catechu) fiber in

polypropylene composite. Journal of Polymers and the Environment. 2012;**20**:501-506

[47] Dhanalakshmi S, Ramadevi P, Basavaraju B. A study of the effect of chemical treatments on areca fiber reinforced polypropylene composite properties. Science and Engineering of Composite Materials. 2017;**24**:501-520

[48] Borah J, Dutta N. Development and properties evaluation of betel nut fibers composite material. Materials Today. 2018;**5**:2229-2233

[49] Chethan M, Krishna SG, Chennakeshava R, Manjunath M. Study on 3-point bending analysis of untreated natural Areca sheath fiber reinforced polymer matrix bio-composites. International Journal of Engineering Research and Advancement Technology. 2016;**2**:338-393

[50] Yousif BF, Devadas A, Yusaf TF. Adhesive wear and frictional behavior of multilayered polyester composite based on betelnut fiber mats under wet contact conditions. Surface Review and Letters. 2009;**16**:407-414

[51] Padmaraj N, Keni LG, Chetan K, Shetty M. Mechanical characterization of Areca husk-coir fiber reinforced hybrid composites. Material Today: Proceedings. 2018;**5**:1292-1297

[52] Prabakaran J, Santhosh S, Saravanakumar C, Rabins RS. Investigation of mechanical properties in Areca (betel nut) and sisal fiber with epoxy composite. Journal for Research. 2017;**3**:23-27

[53] Gupta MK, Srivastava RK. Mechanical properties of hybrid fibers-reinforced polymer composite: A review. Polymer-Plastic Technology Engineering. 2016;**55**:626-642

[54] Jawaid M, Abdul Khalil HPS, Abu Bakar A, Noorunnisa KP. Chemical resistance, void content and tensile properties of oil palm/jute fiber reinforced polymer hybrid composites. Material & Design. 2011;**32**:1014-1019

[55] Jothibasu S, Mohanamurugan S, Vinod A. Influence of chemical treatments on the mechanical characteristics of areca sheath-flax fibers based epoxy composites. Rasayan Journal of Chemistry. 2018;**11**:1255-1262

[56] Salman SD, Leman Z, Sultan MTH, Ishak MR, Cardona F. Influence of fiber content on mechanical and morphological properties of Woven Kenaf reinforced PVB film produced using a hot press technique. International Journal of Polymer Science. 2016;**2016**:7828451

[57] Loganathan TM, Sultan MTH, et al. Physical, thermal and mechanical properties of areca fiber reinforced polymer composites – An overview. Journal of Bionic Engineering. 2020;**17**:185-205

[58] Jayamania E, et al. Comparative Study of Dielectric Properties of Hybrid Natural Fiber Composites, Procedia Engineering. 2014;**97**:545-554. DOI: 10.1016/j.proeng.2014.12.280

[59] Girisha C, et al. Mechanical Performance Of Natural Fiber-Reinforced Epoxy-Hybrid Composites. International Journal of Engineering Research and Applications (IJERA). 2012;**2**(5):615-619

[60] Binoj JS, et al. Morphological, Physical, Mechanical, Chemical and Thermal Characterization of Sustainable Indian Areca Fruit Husk Fibers (Areca Catechu L.) as Potential Alternate for Hazardous Synthetic Fibers. Journal of Bionic Engineering. 2016;**13**:156-165. DOI: 10.1016/S1672-6529(14)60170-0

[61] Yusriaha L, et al. Exploring the potential of Betel Nut Husk fiber as Reinforcement in Polymer Composites: Effect of Fiber Maturity, Procedia Chemistry. 2012;**4**:87-94. DOI: 10.1016/j.proche.2012.06.013

[62] Ashok Kumar M, et al. Materials Science: An Indian Journal. 2011;**7**:219-226 ISSN: 0974-7486

[63] Naik K, Swamy RP. Mechanical behavior of areca fiber and maize powder hybrid composites. International Journal of Engineering Research and Applications. 2014;**4**(185-189):2248-9622

[64] Chethan MR, et al. Study on Tensile Analysis of Untreated Chopped Natural Areca Sheath Fiber Reinforced Polymer Matrix Bio-Composites, International Journal of Engineering Research and Advanced Technology (IJERAT). 2016;**02**:388-393. ISSN: 2454-6135

[65] Dhanalakshmi S, et al. Effect of chemical treatments on tensile strength og areca fiber reinforced natural rubber composites. IOSR Journal of Applied Chemistry (IOSR-JAC). 2015;**8**:43-52-18, e-ISSN: 2278-5736

[66] Srinivasa CV, et al. Flexural behaviour of areca fibers composites, BioResources. 2010;**5**:1846-1858

[67] Srinivasa CV, Bharath KN. Effect of Alkali Treatment on Impact Behavior of Areca Fibers Reinforced Polymer Composites. International Journal of Materials Engineering. 2013;**7**:875-879

[68] Srinivasa CV, Bharath KN. Impact and Hardness properties of areca fiber epoxy reinforced composites. Journal of Materials and Environmental Science. 2011;**2**:351-356. ISSN: 2028-2508

[69] Nirmal U, et al. Betelnut fibres as an alternative to glass fibres to reinforce thermoset composites: A comparative study. Textile Research Journal. 2017;**82**:1107-1120. DOI: 10.1177/0040517512439945

[70] Siva I, et al. Effect of fiber volume fraction on the mechanical properties of coconut sheath/usp omposite. Journal of Manufacturing Engineering. 2013;**8**(1):060-063

[71] Chethan MR, Krishna SGG. Effect of fiber length on tensile characteristics of untreated natural areca sheath FRP green composites. International Journal of Mechanical and Production Engineering. Special Issue. Sep 2016. pp. 90-92. IJMPE-IRAJ-DOIONLINE-5545

[72] Venkateshappa SC, Jayadevappa SY, Puttiah PKW. Mechanical behaviour of Areca fiber reinforced epoxy composites. Advances in Polymer Technology. 2012;**31**(4):319-330

[73] Venkateshappa SC, Bennehalli B, Kenchappa MG, Ranganagowda RPG. Flexural behaviour of areca fibers composites. BioResources. **5**(3):1846-1858

[74] Srinivasa CV, Arifulla A, Goutham N, Santhosh T, Jaeethendra HJ, Ravikumar RB, et al. Static bending and impact behaviour of areca fibers composites. Materials and Design. 2011;**32, (4)**:2469-2475

[75] Mini KM, Lakshmanan M, Mathew L, Mukundan M. Effect of fiber volume fraction on fatigue behaviour of glass fiber reinforced composite. Fatigue & Fracture of Engineering Materials & Structures. 2012;**35**(12):1160-1166

[76] George G, et al. Novel bio-commingled composites based on jute/polypropylene yarns: Effect of chemical treatments on the mechanical properties. Composites PartA: Applied Science

and Manufacturing. 2012;**43**:219-230
DOI: 10.1016/j.compositesa.2011.10.011

[77] Kishore D, Harish R, Allan D J, Ramu M, Sathishkumar J. Development and characterization of areca fiber reinforced polymer composite. Materials Today: Proceedings. 2019;**18**:934-940